LOW COST EMERGENCY WATER PURIFICATION TECHNOLOGIES

Integrated Water Security Series

LOW COST EMERGENCY WATER PURIFICATION TECHNOLOGIES

Integrated Water Security Series

CHITTARANJAN RAY, PH.D., P.E.
RAVI JAIN, PH.D., P.E.

AMSTERDAM • BOSTON • HEIDELBERG • LONDON
NEW YORK • OXFORD • PARIS • SAN DIEGO
SAN FRANCISCO • SINGAPORE • SYDNEY • TOKYO

Butterworth-Heinemann is an imprint of Elsevier

ELSEVIER

Acquiring Editor: Ken McCombs
Development Editor: Jeff Freeland
Project Manager: Priya Kumaraguruparan
Designer: Russell Purdy

Butterworth-Heinemann is an imprint of Elsevier
225 Wyman Street, Waltham, MA 02451, USA
The Boulevard, Langford Lane, Kidlington, Oxford OX5 1GB UK

Library of Congress Cataloging-in-Publication Data
Ray, Chittaranjan.
  Low cost emergency water purification technologies / Chittaranjan Ray, Ph.D., P.E., Ravi Jain, Ph.D., P.E. – 1st edition.
        pages cm – (Integrated water security series)
    Includes bibliographical references
  1. Drinking water–Purification. 2. Water–Purification. 3. Water filters. 4. Emergency water supply. 5. Survival and emergency equipment. I. Jain, Ravi. II. Title.

  TD430.R39 2014
  628.1'62–dc23   2014001187

British Library Cataloging-in-Publication Data
A catalogue record for this book is available from the British Library

ISBN: 978-0-12-411465-4

For information on all Butterworth-Heinemann publications
visit our web site at store.elsevier.com

CONTENTS

About the Authors   vii
Preface   ix

1. Introduction   1
1.1 Standards for Water Quality and Quantity   3
1.2 Technology Requirements   4
1.3 Challenges in Providing Water Treatment for Disaster Relief   5
1.4 Costs   11

2. Technologies for Short-Term Applications   19
2.1 Introduction   19
2.2 High-Energy Systems   19
2.3 Low-Energy Applications   26

3. Solar Pasteurization   31
3.1 Microbiology of Water Pasteurization   31
3.2 Use of Solar Cookers for Drinking Water Production   33
3.3 Devices Designed Specifically for Water   36
3.4 Simple Devices from Common Materials   38
3.5 Commercial Devices in Production   41
3.6 Devices with Recovery Heat Exchange   41
3.7 Water Pasteurization Indicators   44
3.8 Multi-use Systems   47
3.9 The Greenhouse Effect   47
3.10 Use of SOPAS in Conjunction with SODIS   49
3.11 SODIS and Titanium Dioxide   49
3.12 SOPAS and SODIS Technology Evaluation   50

4. Disinfection Systems   55
4.1 UV Light Systems   55
4.2 Silver-Impregnated Activated Carbon   74
4.3 Electrochlorination Systems   79
4.4 Chlorinators   83

5. Technologies for Long-Term Applications 87
  5.1 Slow Sand Filtration 88
  5.2 Packaged Filtration Units 103
  5.3 Pressurized Filter Units 116
  5.4 Small-Scale Systems 122
  5.5 Natural Filtration 130

6. Emerging Technologies for Emergency Applications 169
  6.1 Nanotechnology 169
  6.2 Renewable Energy 170
  6.3 Iodinated Resins 172

7. Water Infrastructure Development for Resilience 175
  7.1 Need for Water Infrastructure Development 175
  7.2 Infrastructure Improvements for Developed Countries 176
  7.3 Infrastructure Improvements for Developing Countries 177
  7.4 Short-Term Solutions 179
  7.5 A Wholesome Approach to Infrastructure Development 180

References 187
Index 199

Chittaranjan Ray, Ph.D., P.E. is the Director of Nebraska Water Center and Professor of Civil Engineering at the University of Nebraska. Prior to this appointment, he was a professor in the Department of Civil & Environmental Engineering and Interim Director of the Water Resources Research Center at the University of Hawaii in Honolulu, HI. He received his Ph.D. in Civil Engineering from the University of Illinois at Urbana-Champaign and his M.S. in Civil Engineering from Texas Tech University. He served as a staff engineer with the consulting firm Geraghty & Miller, Inc. (now part of Arcadis Geraghty & Miller) in Hackensack, NJ between his two graduate degrees. Additionally, he was employed as a research associate and associate scientist at the Illinois State Water Survey in Champaign, IL. His research interests are in the areas of riverbank filtration for water supply, transport of contaminants in subsurface, and general water quality. He is a registered Professional Civil Engineer in Illinois and fellow of the American Society of Civil Engineers. He has published six books, over 75 peer-reviewed journal papers, and many technical reports. He has made dozens of presentations in the areas of water quality, contaminant transport, and riverbank filtration.

    Ravi K. Jain, Ph.D., P.E. Dean Emeritus, was Dean of the School of Engineering and Computer Science, University of the Pacific, Stockton, California from 2000 to 2013. Prior to this appointment, he has held research, faculty, and administrative positions at the University of Illinois (Urbana-Champaign), Massachusetts Institute of Technology (MIT), and the University of Cincinnati. He has served as Chair of the Environmental Engineering Research Council, American Society of Civil Engineers (ASCE) and is a member of the American Academy of Environmental Engineers, fellow ASCE, and fellow American Association for the Advancement of Science (AAAS). Dr. Jain was the founding Director of the Army Environmental Policy Institute, has directed major research programs for the U.S. Army and has worked in industry and for the California State Department of Water Resources. He has been a Littauer Fellow at Harvard University and a Fellow of Churchill College, Cambridge University. He has published 17 books and more than 150 papers and technical reports.

Access to clean water is of serious concern to the human population. Incredible as it may seem, more than one billion people do not have access to adequate water resources. Access to clean water is, yet again, another hurdle as the available water may not be safe enough for human consumption. Natural disasters such as floods, earthquakes, hurricanes, volcanic eruptions, and tsunamis are known to cause severe human miseries. Additionally, wars and civil conflicts also expose civilians to miseries. Under such circumstances, basic survival requires adequate drinking water resources: a major challenge. Water is also needed for sanitation needs, although that can be achieved with somewhat lower quality water during emergency situations.

Production of potable water and its supply to the affected population is the number one priority in most humanitarian relief missions following disasters. Membrane technology is one of the most common methods of producing potable water. Such treatment systems are typically deployed onboard ships or at water sources (lakes, streams, etc.). The next challenge is to transport the produced potable water to the affected population. In the first phase of disaster response, the military, government agencies, and nongovernmental organizations (NGOs) work jointly to deliver water via helicopters, boats, and trucks (if roads are accessible) to the people for the first few weeks. Subsequently, it is best to set up onsite treatment units until permanent solutions are deployed.

Drinking water security is emerging as a crucial national and international concern due to the vulnerability of the water infrastructure and its importance in protecting human health and the economic wellbeing of the population. In addition to this book, *Low Cost Emergency Water Purification Technologies*, which is meant to serve as a guidance document, another book: *Drinking Water Security for Engineers, Planners, and Managers*, is already in press as a part of the *Integrated Water Security Technologies* book series being published by Elsevier.

In this book, we describe appropriate, yet affordable, technologies for producing potable water for emergency use. This book is meant to be used as a primer by NGOs, government entities, the military, planners, and managers who deal with disaster relief. This book describes low-cost technologies that are available commercially or treatment technologies that can be built with indigenous material for short- and long-term applications.

For short-term applications, energy use is not a big barrier. However, energy use can be a major consideration for systems that are to be deployed for long-term use.

Chapter 1 provides an orientation to current thinking about water treatment for emergency application and guides the reader in the selection of appropriate technologies. Chapter 2 provides a description of systems deployed for short-term applications, primarily for the first few weeks following the disasters. The technologies are further classified as high- and low-energy systems based on their energy need per unit of water produced. Technologies considered for short-term use include reverse osmosis, distillation, and forward osmosis.

Use of solar energy for water pasteurization is the theme of Chapter 3. In situations when chemical oxidants are not available and boiling is not feasible (no power, fuel, or firewood), solar pasteurization may be an effective means of killing microbes present in water. This chapter examines commercial as well as homemade units for their effectiveness.

Disinfection of the produced water is described in Chapter 4. Principles of UV irradiation and commercially available UV units for flow-through and static systems are described. Feasibility of LED-based UV lights for small-scale applications is also discussed. Discussed in this chapter also, is use of silver-impregnated, activated carbon; electro-chlorinators; bleach; and chlorine tablets as disinfectants.

Use of treatment systems over a long period of time at low costs are some of the key objectives for disaster response in developing countries. Chapter 5 examines the effectiveness of slow sand filtration as well as natural (riverbank or lake bank) filtration as means of producing large quintiles of water for community use at low cost. This water may not require significant additional treatment besides disinfection. Various packaged filtration systems are described in this chapter. Other low-cost technologies such as Lifestraw and Chuli are described as well. We also describe various small-scale innovative technologies developed in the last few years. For natural filtration, we show improvements in water quality and provide step-by-step guidance for the design of wells and selection of well screens. Protection of wells from surface contamination is also described.

In Chapter 6, we discuss emerging technologies such as nanotechnology, renewable energy, and iodinated resins for water purification. In the last chapter, Chapter 7, we discuss water infrastructure resilience. Here, we discuss the difference between developed and developing countries and provide guidance on how to improve the resiliency of water infrastructures and provide wholesome and sustainable solutions for crucial potable water supplies.

We would like to express our gratitude to several graduate students and research assistants who worked with the authors in researching and compiling data for this book. They include Natalie Muradian at the School of Engineering & Computer Science, University of the Pacific, and Gabriel El Swaify and Matteo D'Alessio in the Department of Civil and Environmental Engineering, University of Hawaii. Natalie Muradian's hard work and dedication in preparing initial drafts of some of the chapters and reviewing page proofs were invaluable to the completion of this work. We acknowledge and thank Dee Ebbeka of the University of Nebraska for her assistance in preparing illustrations.

Chittaranjan Ray
University of Nebraska, Lincoln, Nebraska, USA
Ravi K. Jain
University of Pacific, Stockton, California, USA

CHAPTER 1

Introduction

Contents

1.1 Standards for Water Quality and Quantity	3
1.2 Technology Requirements	4
1.3 Challenges in Providing Water Treatment for Disaster Relief	5
1.4 Costs	11
1.4.1 How Much Should Recipients Pay?	15
1.4.2 Relative Costs	16

Keywords: Water quality, Water treatment, Costs, Point of use, Internally displaced person

More than one billion people do not have access to safe drinking water sources (Lantagne et al., 2010). Many of them reside in developing countries with very few resources available to them. Simple and low-cost technologies have been developed to provide ways to treat water, ranging from point-of-use (POU) treatment to small-scale (SS) community treatment. During natural disasters, POU and SS technologies offer ways to provide clean and safe drinking water. This guide to emergency water treatment has been developed based on current research, products, and field studies to create an expeditious and easy process for choosing which technology is most appropriate in each emergency situation.

Natural disasters, such as floods, tsunamis, hurricanes, and earthquakes, affect more than 226 million people every year (UNISDR, 2011). The occurrence of these natural disasters has been increasing each year (see Figure 1.1) due to the synergistic effect of climate change is causing more extremes in weather and growing populations are living in areas vulnerable to natural hazards (Lantagne and Clasen, 2009). Developing a guideline for emergency water treatment will become even more important as the number of natural disasters continues to increase.

During disasters, water sources become contaminated with industrial, human, and animal waste from overwhelmed sewage infrastructures or because of poor hygiene practices. Diseases caused by waterborne pathogens

☆"To view the full reference list for the book, click here"

1

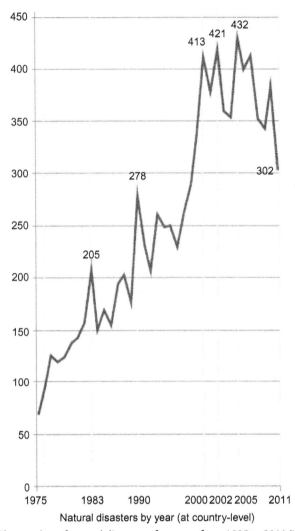

Figure 1.1 The number of natural disasters of per year from 1985 to 2011 (UNISDR, 2012). (For the color version of this figure, the reader is referred to the online version of this chapter.)

can easily be spread by this contamination (Colindres et al., 2007; McLennan et al., 2009), affecting a large number of people, primarily in the developing world. Several billion people use water that is likely contaminated, resulting in approximately 2.5 billion cases of illness per year and about 5 million deaths per year (Burch and Thomas, 1998; Pejack et al., 1996). In the developing world, high rates of morbidity and mortality are caused by common illnesses spread by water contamination, including cholera, typhoid fever, shigellosis, dysentery, and hepatitis A and E (Lantagne and Clasen, 2009). In fact, 40% of deaths in internally displaced person (IDP) camps were due to diarrhea caused by drinking microbiologically unsafe water (Doocy and Burnham, 2006). The communicable diseases present in IDP camps were transmitted through unsafe drinking water. Because of this contamination, water treatment approaches may be valuable in helping to stem a "second wave" of illness and death after a disaster (Colindres et al., 2007).

The poor in developing countries are the most vulnerable to disasters because they do not have the resources to rebuild and fix infrastructures (UNISDR, 2011). After natural disasters, survivors either leave their country and become refugees or stay in the country and migrate to someplace safer, becoming IDPs. Close living quarters, lack of hygiene, and insufficient clean water supplies in IDP camps can exacerbate already poor conditions (Steel et al., 2008). Even if members of an affected community remain at home rather than moving to an IDP camp, there is still a chance that their water will become contaminated.

1.1 STANDARDS FOR WATER QUALITY AND QUANTITY

During a disaster, clean water is necessary for survival. The main health problems associated with drinking water contamination are caused by insufficient water for hygiene purposes and consumption of that contaminated water. There are two standards defined by the Sphere Project (2011) for water supply standards. The first standard involves the quantity of and access to water, while the second standard regulates water quality. The minimum amount of water for safe and healthy consumption is summarized in Table 1.1. Water for hygiene use is considered a basic water need because it is important maintaining proper hygiene during disasters to reduce the risk of disease.

Table 1.1 Basic Survival Water Needs (Sphere Project, 2011)

Water Use	Minimum Requirement (L/day)	Notes
Survival needs: water intake for food and drinking	2.5-3	Depends on the climate and individual physiology
Basic hygiene practices	2-6	Depends on social and cultural norms
Basic cooking needs	3-6	Depends on food type and social and cultural norms
Total	7.5-15 L/day	

Table 1.2 Water Quality Parameters for Disaster Relief (Sphere Project, 2011)

Water Quality Parameter	Minimum Requirement	Notes
Coliforms	0/100 mL of water	Measured at the point of delivery
Turbidity	<5 NTU	For proper disinfection
Chlorine Residual	0.5 mg/L	To reduce risk of posttreatment contamination

Along with the 7.5- to 15-L of water per person per day, other water quantity indicators that can be used to measure the amount of water accessibility include:

- Distance from a household to a water point ($\sim$500 m);
- Waiting time at a water source (no longer than 30 min)

These standards should not be followed blindly, as they do not guarantee that water is equally available to all.

Water quality is a secondary standard, according to the Sphere Project (2011). Once water quantity has been assured, water quality should be improved to reduce the risk of dysentery and other diseases. The quality parameters identified in Table 1.2 are specified by the Sphere Project (2011) as the minimum standards that must be met by water treatment technologies.

1.2 TECHNOLOGY REQUIREMENTS

Because the majority of people affected by natural disasters are not knowledgeable about the technologies available for emergency water purification, criteria for technology use should reflect this. There are two types of responses in natural disasters. The first response is a rapid (and potentially short) response to a disaster, designed to keep people alive. During disasters, this response tends to include the transportation of bottled water to the affected population. However, the weight and expense of transporting bottled water has resulted in a lack of sustainability, so alternatives are continually being explored. The second response is a more sustainable and long-term response. This second response needs to last until the IDP camps or community living conditions return to normal. These technologies can even be appropriate for future water treatment during stable times if water treatment was unavailable before the emergency.

Table 1.3 The Flow Rate Needed for a Water Treatment Device Based on How Many People Are Being Served

Persons Per Water Source	Treatment Capacity Needed (L/h)
1	1
5	5
25	24
50	47
100	94
250	235
400	375
500	469

Initial, rapid response for water treatment should have the following characteristics (Ray et al., 2012):
• Provides needed quantity of water, and the water is of drinkable quality
• Portable
• Low cost
• Light weight
• Ease of use or requires minimal training
• Requires minimal or no external power

A solution for a long-term response should have the following characteristics (Ray et al., 2012):
• Ability to support a community or large population by purifying a large volume of water
• Parts require infrequent replacements and minimal supply chains
• Does not require complex training to operate
• Uses easily available power sources

Water treatment technologies available in an emergency must provide the minimum amount of water needed for basic survival; thus, the required capacity of these technologies depends on the amount of people the device serves. Using the minimum amount of water needed for survival from Table 1.1 (7.5 L/day) and assuming that water can be collected for 8 h a day (Sphere Project, 2011), the required capacity of water treatment technologies can be estimated; see Table 1.3.

1.3 CHALLENGES IN PROVIDING WATER TREATMENT FOR DISASTER RELIEF

Water treatment technologies are typically provided to those in need through nongovernment organizations (NGOs), the government, or companies that manufacture water treatment devices. The effectiveness of these organizations in responding with water treatment technologies to these disasters has not been extensively studied in peer-reviewed journals. However, a study by Colindres et al. (2007) did find that NGOs were less efficient then they could be. When delivering water treatment products, NGO representatives were not familiar with the treatment process they were distributing, nor did they have the training or experience needed to promote the water treatment device (Colindres et al., 2007). In order to provide the most help possible, the organization that is offering relief technologies needs to be familiar with the product they are distributing and be trained in using the technology. Emergency responders should include local community leaders in the training, educational, and distribution process to gain community support for the technology. This will also help the local population to understand that licensed technologies provide a way to help them stay healthy and can offer alternatives to drinking dirty water (Colindres et al., 2007).

A difficulty in measuring the effectiveness of emergency water treatment technologies is that metrics are difficult to develop and compare in different emergency situations. For example, the majority of studies that evaluated disaster relief programs were all conducted during a stable time after the emergency, not during the initial, acute response. Thus, the technologies that were found to be effective through field tests and applications may not be as easily implemented in the more chaotic situation of an acute emergency response (Colindres et al., 2007). Another difficulty in assessing the effectiveness of water treatment technologies in an emergency situation is the difference in local conditions. One study found that using chlorinated tablets reduced dysentery in a community while another study found that the same chlorinated tablets did not reduce dysentery in a different population (Clasen et al., 2007; Jain et al., 2010). The numerous factors related to water quality, including the amount of microbiological activity in the water, water storage practices, hygiene, and water treatment technology education, decrease researchers' ability to evaluate and compare results.

Studies that have researched emergency water treatment found that the greatest challenges during implementation were user acceptance and user training (Lantagne and Clasen, 2009). Due to the chaos of acute emergency situations, user training can be difficult to implement. Lantagne and Clasen (2009) found that initial stages of emergency situations had less than a 20% uptake of POU water treatment devices. More stable emergency situations resulted in a higher rate of POU uptake due to the implementation of training. Training is important because effective treatment of water depends on correct use of the device. If the device is not used correctly, users will continue to get sick and assume that the device does not actually work, discontinuing its use. This fosters a lack of trust in water treatment in general. User acceptability is also dependent not only on correct use of the device, but ease of use, proper maintenance, and taste of the treated water.

Implementing emergency devices in a manner that reduces illnesses and is sustainable is difficult because each disaster affects different communities in different ways. Some emergency devices that may work well in some areas may not be accepted in other regions. For example, some regions of developing countries already use a certain type of water treatment during stable times, so implementing the same type of treatment might be more effective than implementing a new technology. Loo et al. (2012) developed a procedure that can be followed during an emergency situation. While this procedure is new and has not yet been tested in an emergency, it provides a guideline to follow that would help implement the most effective technology depending on the scenario. The list of parameters to consider when following the procedure is provided in Figure 1.2. After determining the important parameters, the decision tree in Figure 1.3 can help determine the technologies that are the most appropriate. After selecting several possible technologies, the matrix designed by Loo et al. (2012) in Figure 1.4 can be used to compare the technologies and see which technology is the most appropriate. The matrix in Figure 1.4 can also be tailored to fit the site-specific needs. Equation 1 can be used to determine the nominal and weighted score of each potential water treatment technology listed in Figure 1.4. Than that score can be applied in Step 4 of Figure 1.2.

Score of water treatment device $= \sum_{j-1}^{n} w_j x_{ij}$ (Equation 1, Loo et al., 2012)

Where wj = weighting factor arbitrarily applied to criteria (See top row in Figure 1.4)

Xij = score for each criterion (See columns in Figure 1.4)

When implementing a new water treatment technology, factors to consider include:

- Water source and quality
  - Surface water, types of wells, spring water, pipe network
    - If not piped, how water is collected
  - Turbidity, microbiological activity, arsenic
  - Potential contamination: proximity of nearby industry/agricultural areas and locations of community sewage to water source
- Initial, acute response or long-term application
  - Recipient's culture

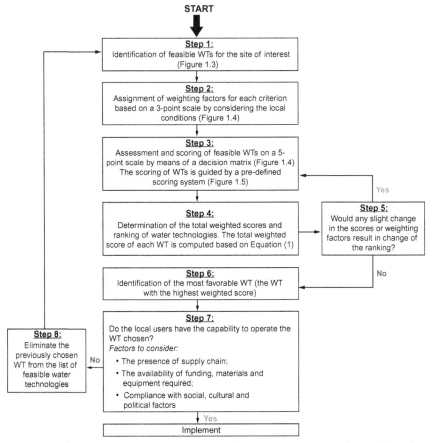

Figure 1.2 Flowchart for the water treatment selection process (Loo et al., 2012). (For the color version of this figure, the reader is referred to the online version of this chapter.)

- Recipient's familiarity with similar technology and/or previous technology use
- Recipient's hygiene practices
- Previous training and adaptability
- How and what water was used for and how it will be used now
- How water is stored and how recontamination of treated water can be prevented
- Cost: Initial investment and maintenance costs
  - Lifecycle costs; see Section 1.4
- Operation and maintenance
  - Labor required to operate and clean device on daily basis
  - Availability of replacement parts

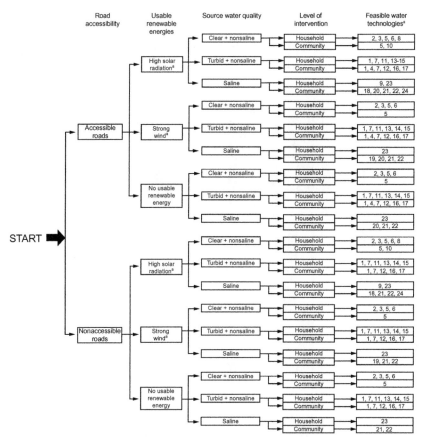

Road accessibility	Usable renewable energies	Source water quality	Level of intervention	Feasible water technologies[a]

Figure 1.3 Decision tree for choosing the most appropriate water treatment technology (Loo et al., 2012).

Note: 1 = Biosand filter; 2 = Boiling; 3 = *Chulli* purifier/WADIS; 4 = Land-based mobile water treatment plant; 5 = NADCC tablets; 6 = UV disinfection (portable); 7 = PUR® sachet; 8 = SODIS; 9 = Solar still; 10 = Solar water heater; 11 = Structured matrix filter; 12 = Upflow clarifier; 13 = Household ceramic filter; 14 = Portable MF; 15 = Portable UF; 16 = Modular UF; 17 = Bicycle-powered NF; 18 = Modular PV-RO; 19 = Modular wind-powered RO; 20 = Land-based mobile RO plant (generator); 21 = Mobile floating RO (generator); 22 = Modular RO (generator); 23 = FO filter pouch; 24 = Compact SMADES (MD powered by solar energy).
[a]The list of feasible water technologies also includes those powered by nonrenewable energy such as electrical generator, fuel, diesel, etc.

- Power sources available
  - Renewable energies: solar radiation, wind energy
  - Car batteries
  - Electricity

Using the above selection strategy requires detailed information about the disaster area. This information may not be known prior to selecting a treatment technology. Steele et al., (2008) recognize that there will be a lack of sufficient knowledge when it comes to the decision-making process and recommend addressing this issue by thoroughly documenting the advantages and disadvantages of each water treatment technology. When the time

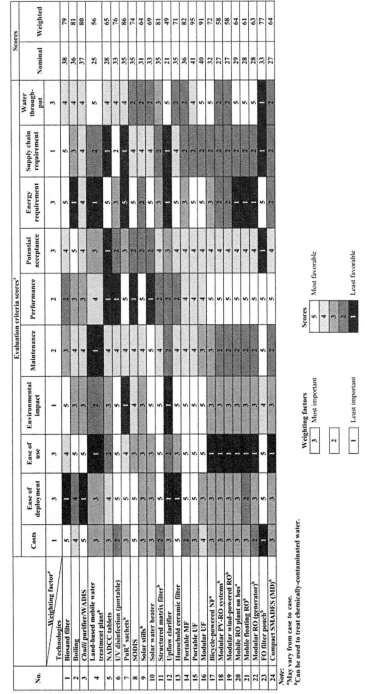

Figure 1.4 Matrix used to compare emergency water treatment technologies (Loo et al., 2012). (For color version of this figure, the reader is referred to the online version of this chapter.)

comes to evaluate the needs of the disaster site, the treatment technology can be chosen based on available knowledge (Table 1.4).

1.4 COSTS

When calculating costs for emergency relief water treatment technologies, all costs that will be incurred throughout the lifetime of the device must be taken into account to assess the device's appropriateness. Cost considerations include upfront, recurring, and program costs. Upfront costs must include the initial cost of the hardware as manufactured, transportation, and installation costs. Installation costs can depend on who is doing the installation, locals or volunteers from a humanitarian program. Some devices may not even have installation costs or may have training costs instead. Recurring costs must include the daily labor needed for operation and maintenance and the costs of repairing and replacing any broken or old parts over the device's lifetime (Burch and Thomas, 1998). Accurate estimates of maintenance and labor costs may be difficult to calculate. Total costs should also include program costs that make the implementation of the device successful, such as education and training.

Costs can be normalized in several ways. The typical way to define cost is the life cycle cost per unit volume of water produced; however, cost can also be defined as the capacity cost, which is the initial cost per unit of daily capacity or the amount of water treated per month for a household. As long as the same indicator of normalization is used, costs can be compared for many different devices. For example, a treatment system that uses membranes may have higher upfront cost than a candle filter system, but, because the membrane system has a longer lifespan and a greater flow than a candle filter, the normalized costs may be comparable.

Programs that provide humanitarian relief during disasters, such as AmeriCares, PSI, UNICEF, World Vision, International Rescue Committee, International Federation of Red Cross, GWI, USAID, and Samaritan's Purse, all operate differently (Lantagne et al., 2007). This means their costs for products will all be different. Expenses vary by location and distribution strategy, that is, how the device is transported and installed and whether training/demonstrations are made available. Studies have shown that programs that incorporate training methods into treatment device distribution have a higher rate of social acceptance of the device, and, thus, the device is more effective. Working closely with local governments, providing technical assistance, and training locals to become technical experts are program additions that increase the effectiveness of the program but also increase the cost of the program.

Table 1.4 Criteria Used to Evaluate Water Treatment Technologies and Descriptions of Scores (Loo et al, 2012)

Evaluation criteria	Definition of Scores				
	1	2	3	4	5
Costs	Very high cost per liter (>1.00 US $/L)	Moderate cost per liter (0.10–100 US $/L)	Low cost per liter (0.01–0.10 US $/L)	Low cost per liter (<0.01 US$/L)	Only one-time cost is required (<10 US$/unit)
Ease of deployment	Large and heavy; require construction and assembly of the whole system onsite	Large and heavy; require relatively extensive assembly of parts of the system	Moderately large and heavy; require some simple set-up of the WT	Light and small sized; require some simple household materials for set-up	Light and small sized; no set-up required
Ease of use	Very complicated process design; can only be operated by skilled operator	Difficult to be operated by un skilled personnel; require determination of proper dosage of chemicals	Require some simple training to user; long treatment time (>1 h)	Simple training is required to start using; short treatment time (<1 h)	Essentially no training required to start using short treatment time (<1 h)
Environmental impact	Produces environmentally malign byproducts; toxic; large quantity	Produce environmentally malign byproducts; toxic; small quantity	Produce environmentally malign byproducts; mild effect; small quantity	Does not produce any environmentally malign byproducts; use non-biodegradable materials	Does not produce any environmentally malign byproducts; does not use any non-biodegradable materials

Maintenance	Complicated maintenance; done regularly; time-consuming	Complicated maintenance; done regularly; not time-consuming	Slightly complicated activities; done regularly; not time-consuming	Simple maintenance; done occasionally; not time-consuming	No maintenance required
Performance	Modest microbes removal; treatment performance is affected by variations in source water quality; cannot remove turbidity	Modest microbes removal; treatment performance is affected by variations in source water quality; can remove turbidity	Excellent microbes removal; treatment performance is not affected by variations in source water quality; cannot remove turbidity	Excellent microbes removal; treatment performance is not affected by variations in source water quality; can remove turbidity	Excellent microbes removal; treatment performance is not affected by variations in source water quality; can remove a wide range of contaminants (either chemicals or salt)
Potential acceptance	No visual improvement of treated water; treated water may have objectionable taste; may produce harmful byproducts or	No visual improvement of treated water; no objectionable taste	Visual improvement of created water; involve addition of chemicals into water which may not be acceptable to some users; no	Visual improvement of treated water; does not result in objectionable taste; does not produce harmful byproducts	Common practice among users

Continued

Table 1.4 Criteria Used to Evaluate Water Treatment Technologies and Descriptions of Scores (Loo et al., 2012)—cont'd

Evaluation criteria	Definition of Scores				
	1	2	3	4	5
	does not produce pure water but a sweetened drink		objectionable taste		
Energy requirement	Uses large amount of energy and cannot be powered by renewable energy	Uses a large amount of energy but can be powered by renewable energy	Can be powered by small hand pump or bicycle	Require energy/ fuel for operation but do not involve additional use of energy	No power requirement (gravity fed or mouth suction)
Supply chain requirement	Require continuous supply of consumables; consumables are only available from specific vendors	Periodic replacement of damaged parts; replacement parts are only available from specific vendors	Require continuous supply of consumables; consumables are off-the-shelf materials	Periodic replacement of damaged parts; uses off-the-shelf materials	No supply chain required
Water throughput	Very low yield (<3 L/day)	Low yield; depends on meteorological conditions	Moderate yield	High yield; can serve a small community of people or household	High yield; can be used to serve a large community of people

Because programs affect the variability of cost so much, it might be appropriate to examine costs of water treatment technologies without considering program costs such as training. However, some devices can be implemented with less training than others and thus might be more cost-effective in certain situations. In fact, the training associated with devices is what prompts effective water treatment. Program costs associated with each device need to be considered. In order to accurately assess and compare costs, a template could be created with cost estimates for:

- Manufactured cost of hardware and lifetime of device
- Transportation: how far and by what method the device will be transported from manufacturer to consumer
- What installation is needed, how long it takes to install, and who is required to install
- Daily labor costs involved could include:
  - The cost of paying someone to operate the system, or
  - Opportunities lost elsewhere by operating the device
- How often parts need to be replaced and repaired, taking into account supply chain availability
- Education: training locals, training local safety officials, educating local governments, performing demonstrations

Several studies have mentioned that the cost of the initial response to provide drinking water during an emergency is somewhat disregarded (Lantagne and Clasen, 2009). Only when the first, rapid response transitions to a more long-term treatment approach does cost become more of a consideration. Making the change from the high-cost treatment used in a rapid response to low-cost, sustainable treatment options helps save money. Programs that strive to implement sustainable practices face financial challenges as well because longer programs require more staffing and a supply chain for replacement parts. However, implementation of these sustainable programs is possible. As an example, Potters for Peace, a program that teaches pottery to locals, was initially funded by private donations but is now a self-financed micro-enterprise (Lantagne et al., 2007).

1.4.1 How Much Should Recipients Pay?

Several groups of researchers conducting studies distributed a water treatment product for free in developing countries (Clasen and Boisson, 2006; Colindres et al., 2007; Doocy and Burnham, 2006; Lantagne and Clasen, 2009) and evaluated the willingness of locals to pay for the product after the study was over (Colindres et al., 2007; Lantagne and Clasen, 2009). These studies found that the willingness to pay for the products was always less than what the products would be sold for at average product cost (not including transportation, distribution, and marketing). Even in cases where the products were not distributed for free, people's willingness to pay for the technology was almost always lower than the selling price mainly because of their very low incomes. This simply means that clean and safe drinking water needs to be heavily subsidized. For water treatment devices to be used and maintained effectively, the recipients must have a desire to use the device as well as pay for at least a small part of the true cost of the device (Burch and Thomas, 1998). This generates ownership of the device and ensures someone will be responsible for the upkeep of the device. Studies have shown that people who owned a water treatment device were more likely to use their device if they sought out the device, rather than being offered the device by nongovernmental organizations (Gupta et al., 2008).

Different NGOs and programs offer different ways of providing subsidized water treatment to communities in the aftermath of emergencies. Some programs sponsor the hardware needed for safe water storage and sell a chemical disinfectant at a low cost. Some programs provide a filtration device and require that the users transport and construct the device and become trained in how to use it. Another program installs the filters, provides training, and additionally trains local officials to promote correct use of the device. These different relief efforts all have different costs associated with their programs and treatment devices but cannot charge the true cost of the device to the consumers as they are unable to pay. Donations from faith-based organizations, corporations/agencies, and private donations to various relief funds bridge the gap between what the recipients can afford and what the devices actually cost (Lantagne et al., 2007).

1.4.2 Relative Costs

When implementing water treatment systems, it is important to consider that POU water treatment is more effective in preventing disease and costs less than installing protected wells and springs (Clasen and Boisson, 2006). The costs of implementing POU alternative emergency water treatment need to be compared to costs of either prior treatments used (such as the aforementioned wells or boiling water) or the costs of not using any treatment (medicine, opportunities lost due to illness, etc.). Alternative water treatment devices may seem expensive initially, but, comparatively, the cost of using these devices ends up being much less. Discrepancies exist between the different definitions of "inexpensive" in these studies, however. Some devices are described as inexpensive when, in reality, they are expensive for recipients (Gupta et al., 2008; Islam and Johnston, 2006). These discrepancies could be caused by the studies' authors' ignorance of the percentage of income recipients can afford to spend on water treatment devices.

CHAPTER 2

Technologies for Short-Term Applications

Contents

2.1 Introduction	19
2.2 High-Energy Systems	19
2.2.1 Reverse Osmosis	21
2.2.2 Distillation Technology	24
2.3 Low-Energy Applications	26
2.3.1 Forward Osmosis	27
2.3.2 Emergency Use	27
2.3.3 Commercial Products	29
2.3.4 Costs	29
2.3.5 Evaluation	29

Keywords: Energy, Electrodialysis, Membrane distillation, Forward osmosis, Reverse osmosis

2.1 INTRODUCTION

For short-term applications, water treatment technologies usually need to be implemented with short notice, such as in humanitarian assistance/disaster relief (HA/DR) scenarios following natural disasters or in United Nations-operated refugee camps during wars or armed conflicts. The main concern during short term emergency response is to reduce mortality rates. As a result, energy is not considered a constraint. In addition, water can come from any reasonable source. For example, if the HA/DR site is located on a coast, sea water is desalinated on board ships then transported to consumption centers. Shipboard power systems or auxiliary generators are used to power the water treatment units.

2.2 HIGH-ENERGY SYSTEMS

For short-term applications, water treatment technologies usually need to be implemented with short notice, such as in HA/DR scenarios following

---

☆"To view the full reference list for the book, click here"

Low Cost Emergency Water Purification Technologies

natural disasters or in United Nations-operated refugee camps during wars or armed conflicts. In such scenarios, the water can come from any reasonable source in the vicinity of where the treatment units are installed. If the HA/DR site is located on a coast, sea water is desalinated on board ships and then transported to the consumption centers. In such short-term applications, energy is not considered as a constraint. Shipboard power systems or auxiliary generators are used to power the water treatment units.

It is well known that membrane filtration, and distillation are some of the most expensive treatment technologies available. Membrane filtration is divided into two major groups (Randtke and Horsley, 2012): pressure-driven membrane processes such as reverse osmosis (RO), nanofiltration, ultrafiltration (UF), and microfiltration (MF); and voltage-driven membrane processes such as electrodialysis (ED) and electrodialysis reversal (EDR). While ED and RO are both used for desalination, ED removes only ionic contaminants and is ineffective in removing uncharged species. RO is the most common pressure-driven membrane filtration process for HA/DR applications. Table 2.1 provides a summary of three technologies in removing various contaminants present in the source water.

RO units produce only a fraction of the source water and use a lot of energy. However, RO is better than distillation, where energy use can be higher per unit produced. For RO, the source water can come from a low-quality surface source, brackish ground/surface water, or sea water. The energy needs for large commercial RO units increase with the increase of source water salinity, as more pressure is needed to push salt water through RO membranes.

Table 2.1 Removal Efficiencies of Three Treatment Technologies for Common Contaminants

	Chloride	Sulfate	Nitrate	Pesticides	Bacteria	Protozoa	Viruses	Chlorine	Mercury	Arsenic	Lead	Taste/Odor	VOCs	Fluoride
RO	+	+	+	+	++	++	+	+	+	+	+	−	+	−
Carbon block filters	−	−	−	++	+	+	−	−	−	−	+	+++	++	−
Water distillers	++	++	++	+	+++	+++	+++	++	++	++	++	+	+	++

A large amount of sea water needs to be boiled and the steam condensed to make potable water. Disposal of salt in the reject water is also a concern. In addition, distillation units must be cleaned of the deposited salt after boiling sea water. Multistage flash (MSF), multiple effect flash, vapor compression, and membrane distillation are some of the distillation processes used in commercial units. The large distillation units require operators with a lot of technical competence, and they cost a lot of money. Furthermore, it takes a significant amount of time to construct these large units, rendering their use limited in HA/DR applications. However, smaller distillation units are available for home use that can also be used during emergency situations.

ED/EDR units, typically used for fresh water systems, produce nearly the same amount of potable water from the fresh water. Although these units are less energy intensive, the unit production costs can be high because these systems use expensive filters that eventually need to be replaced. Almas (2006) reported that the cost of a home distillation unit per gallon of water is $0.37/gal; in comparison, an RO device costs about $0.16/gal (Culligan brand). These costs are relatively high. ED and EDR units produce effluent in nearly the same amounts as the feed water; however, they are quite complicated, and poor removal of organic uncharged compounds can be a concern when source water quality is poor.

2.2.1 Reverse Osmosis

Reverse osmosis (RO) is, by far, the most common high energy–using system deployed during HA/DR scenarios. Saline water is pressed through semipermeable membranes at higher than osmotic pressure to produce drinking water (Figure 2.1). The osmotic pressure increases with water salinity. Large RO systems operating at pressures exceeding the osmotic pressures are used to ensure an acceptable rate of water production. Normally, the water production for RO membranes is reported in gallons of filtrate per square foot of membrane surface area per day (gfd) or in other units such as liters of filtrate per square meter of membrane surface area per hour (LMH). Alcolea et al. (2009) estimated that global membrane desalination will reach 97.5 million m^3/day in 2015. Rajagopalan (2011) notes that the U.S. desalination market should reach about 11 million m^3/day at the same time. He states that major regions where new capacities are expected to be added include North Africa, the Middle East, East Asia, the Pacific, and parts of Western Europe.

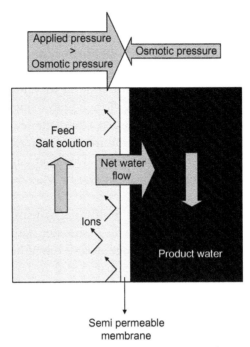

Figure 2.1 Schematic diagram of reverse osmosis process (after Rajagopalan, 2011).

The U.S. Navy has been using shipboard RO units to desalinate seawater and helicopters or other means to deliver water to the affected population. There are also many commercial companies that offer shipboard desalination for supplying water to shoreline cities. Below are some examples:

1. Water Standard Company (www.Waterstandard.com):

    Water Standard Company specializes in onshore and offshore water treatment for oil and gas operations in the ocean. Recently, the company awarded an engineering, procurement, and construction contract to Veolia for fitting its ship *H2Ocean Cristina* with a seawater RO desalination facility to produce a minimum of 13 million gallons/day of potable water.

2. ACWA Power International (www.acwapower.com):

    ACWA Power International has business lines on power production and desalination. The company produces 2.3 million m^3/day of water (more than 600 million gallons/day) from desalination and has an electric generation capacity of nearly 16,000 MW. For the City of Jeddah, Saudi Arabia, they developed two barges, each producing 25,000 m^3/day (6.6 million gallons/day) of potable water.

The above are extremely large-scale RO projects and primarily use seawater as the source water. However, there are other, land-based commercial units that the military has been using during HA/DR missions for fresh or brackish or even seawater sources. Aspen Water (www.aspenwater.com) is one such company that has been supplying units to the military for HA/DR use. Aspen 2000DM is a compact unit placed in a pelican case that can also be pulled (see Figure 2.2). The unit requires 1500 W of power and can operate on AC and DC power sources. The pump in the system produces up to 1200 lb/in.^{2} (psi) pressure. Besides the prefiltration unit(s), it comes with an activated carbon postfiltration unit to remove taste and odor problems. A UV disinfection unit is also attached to the unit for microbial disinfection or virus inactivation. The unit weighs about 400 lb.

Aspen has a number of other units for various source waters. Aspen 1000DM can desalinate up to 1000 gallons/day from any water source including seawater. It can run on single-phase AC or even from 24-V DC power from military vehicles. The power requirement is about 650 W. Solar panels can also be used to charge a bank of batteries that

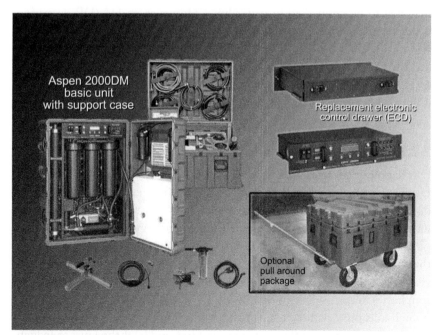

Figure 2.2 Aspen Water's 2000DM unit that produces about 2000 gallons of water per day (see http://www.aspenwater.com/sitebuildercontent/sitebuilderpictures/2000DM_collage_1-18-10.jpg).

can run the unit. The unit comes with conventional RO followed by UV light for sterilization. The unit weighs about 220 lb.

Aspen 3300 M, 5500 M, and 10,000 M are designed to use fresh water sources and the membrane pressures vary between 20 and 45 psi. The units are designed to produce 3300, 5500, and 10,000 gallons/day of product water using a fresh water source. These units are ideal in HA/DR situations when the water source is a pond or stream. The units require two-stage pre-filtration prior to RO. Each of these units is equipped with one UV unit for microbial disinfection. The weight of these units varies depending on the model. The 5500 M unit weighs around 200 lb and can operate on single-phase AC or batteries.

The costs of the Aspen units range upward of $10,000. Most of the units are modular, and parts can be ordered through Aspen or directly from the manufacturers. Most recently, Aspen has worked with a U.S. solar panel maker (Greenpath Technology, see www.greenpath-tech.com) for using their foldable solar panels for powering the fresh-water RO systems.

Home, or small community-scale, systems for fresh water are easily available in the market from dozens of manufacturers. Many of these systems are designed to be connected to water lines. However, in HA/DR scenarios, water lines are typically nonfunctional. External pumps and storage tanks will be needed to supply water to the unit and to store feed water. Costs for the units without tanks and external pumps/generators are about $5,000 for a daily flow rate of 4,000–5,000 gallons. Adding the costs of pumps, generators, and tanks can increase the price to $8,000 or more.

2.2.2 Distillation Technology

Distillation of brackish or sea water or poor quality fresh water is possible. Distillation produces water in its purest form. Numerous vendors produce distillation units for home drinking water, but distillation is a slow and energy-intensive process. Different technologies are available depending on the scale of the application. Most home distillation units produce no more than a few gallons per day although some produce as much as 10 or more gallons per day. The processes involved in home distillation units are very simple: (a) water is first boiled inside the distiller using electrical energy (which kills most pathogens), (b) water vapor is produced from the boiled water, leaving salts and other impurities behind, and (c) the steam is then condensed and water droplets are collected in a clean container. During the vaporization process, it is possible that some volatile contaminants may also be vaporized. If the heat does not destroy the chemicals, then there is a chance that these contaminants will reappear in the condensate (Table 2.1). Table 2.1 provides a summary of three technologies in removing various contaminants present in source water. Distillation has a higher removal efficiency of a wider range of contaminants compared to RO and carbon block filtration.

Small distillation units are suitable for families living adjacent to an ocean, where fresh waters are naturally brackish. Some vendors that produce distillation units for use in the home or in small offices include Durastill, Precision, Pure Water, Polar Bear, West Bend, and StreamPure (see: www.Waterdistillers.com). The water from these units is typically used for drinking and cooking. Small countertop units cost around $500 and can produce a gallon of water in 4–6 h. Larger units can produce 8–12 gallons per day and cost between $2,500 and $4,000. Annual costs for supplies and maintenance vary depending on the source water and the size of the unit and could range from a few hundred dollars up to almost $1,000.

Commercial entities, resorts, and industries in arid places sometimes use sea water distillation to produce potable water. If the distillation units are damaged or destroyed during an emergency, it may be possible to repair them during the recovery phase. The MSF distillation process (Figure 2.3) has been used for quite some time in desert countries in the Middle East such as Saudi Arabia, Kuwait, and the United Arab Emirates. As shown in this figure, incoming sea water is heated in pipes in various stages of the MSF unit and further heated in the heating section of a boiler (at a temperature between 90 and 110 °C). The hot sea water is then returned to the first stage of the MSF, which is maintained at a pressure lower than the equilibrium pressure.

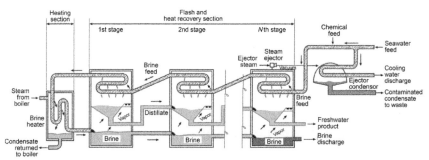

Figure 2.3 FO products from HTI (2010). (a) Expedition; (b) X-Pack; and (c) Hydropack. (For the color version of this figure, the reader is referred to the online version of this chapter.)

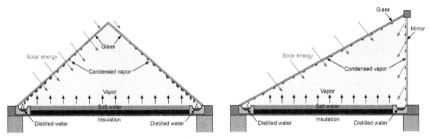

Figure 2.4 Diagram of a multistage flash distillation plant (Buros, 2000; USAID, 1980). (For the color version of this figure, the reader is referred to the online version of this chapter.)

As a result, flash boiling (instantaneous and violent boiling and evaporation) takes place. Fresh water is formed by the condensation of water vapor and is collected in each stage. The pressure is generally lower in the second stage than in the first stage. The pressure has to be low because the brine is cooler and has higher salinity than in the previous stage. In the heat rejection stage, the incoming sea water is used to cool the ejected steam.

Buros (2000) provides some details of desalinating techniques, including thermal distillations (MSF, multiple effect flash, and vapor compression), membranes (ED and RO), and minor processes such as freezing, membrane distillation, and solar humidification. He reported that, for large MSF facilities in Abu Dhabi in the 1990s, the cost to deliver water was between $2.80 and $3.00/1000 gallons. Saidur et al. (2011) conducted a detailed review of different distillation methods for small-scale applications (Figure 2.4). Among the many distillation techniques, solar stills may be useful in situations where the climate is hot and there is abundant sunshine all year round. Reviews of solar stills and their use are widely reported in literature (Dev and Tiwari, 2011; Kaushal and Varun, 2010; Tiwari et al., 2003). In particular, Tiwari and colleagues and Dev and Tiwari conducted energy-balance analyses to estimate the amount of water production by solar stills. Solar stills can be used in the recovery period, especially for long-term applications.

2.3 LOW-ENERGY APPLICATIONS

There are many treatment technologies that do not require as much energy as RO, distillation, ED, and so on. Filtration processes, including GAC filtration, are less energy intensive as they only remove pollutants. Natural filtration processes such as biosand or slow sand filtration and riverbank filtration are extremely low-energy treatment systems. However, they take more time to set up and may not be as useful for very short-term applications. Cartridge or ceramic filtration as well as packaged filtration units are less energy intensive and can be adapted for short-term use.

The cost to implement a small biosand or slow sand filtration unit for home use is typically under $100. Commercial vendors (see www.bluefuturefilters. com) can set up affordable community-scale systems that produce 6,000 to 12,000 gallons per day. These units typically require some type of disinfection technology such as ultraviolet (UV) light, ozone, or chlorine.

2.3.1 Forward Osmosis

Forward osmosis (FO) is the spontaneous flow of a solvent across a membrane. The driving force for this process is the difference in chemical potential between two fluids on either side of a membrane. The solvent in this process is typically water. Untreated water with a high osmotic potential (or low concentration of a solute) is on one side of the membrane. A solution with a low osmotic potential (or a high concentration of a solute) is on the other side of the membrane. Water flows from the low concentration of solute to the high concentration of solute (i.e., from the dirty water to the treated water). FO does not produce clean water, but rather produces a "design solution," a diluted salt or sucrose solution. Membranes used for FO must have pore sizes smaller than viruses and bacteria to allow only the water molecules through the membrane and must be thin enough to allow water to flow through the membrane (Miller and Evans, 2006).

The use of FO to remove bacteria and toxins from water requires no power and does not result in membrane fouling. Therefore, FO technology has several possible applications, such as in desalination plants, RO plants that have difficulty with membrane fouling, irrigation in areas with brackish aquifers, and short term emergency water treatment applications.

2.3.2 Emergency Use

Several mobile packs that use FO to purify water during emergencies have been developed. These packs have a large advantage over solar disinfection, chlorination, and UV lamps because FO can effectively remove pathogens and bacteria from muddy and turbid water (Cath et al., 2006). NASA and the Department of Defense are exploring urine recycling using a variation of this design (Cohen, 2004).

There are several types of FO packs. A reusable device consists of a double-lined pack that has a large outer bag that holds untreated/dirty water. The inner compartment is a bag made out of a special FO membrane that contains a nutrient solution (typically flavored sucrose). The dirty water from the outside pack diffuses through the membrane into the inner compartment. This process is driven by osmotic differences between the source water and the nutrient solution.

A simpler device consists of just a membrane bag that holds a nutrient solution, such as the Hydropack from HTI (2010) (Figure 2.5c). The bag is activated when it is placed into dirty water.

The water diffuses through the membrane, leaving all impurities and toxins behind. The end result is a nutrient-rich drink that can boost energy and provide necessary nutrients that may not be otherwise obtained in a disaster situation (Salter, 2006). While the nutrient-rich drink is advantageous to someone in dire need of water, this solution is not ideal for cooking or hygiene. The maintenance required for these packs is to refill the membrane bag with the nutrient solution after a certain number of liters of water have been treated. In some cases (such as the Hydropack from Hydration Technologies Inc.), the bag can only be used once. Filtering times range from 20 min for the Expedition (Figure 2.5a) to 4-12 h for the X-Pack (Figure 2.5b) and the Hydropack (Figure 2.5c).

Similar to the delivery of water bottles in emergency situations, these packs must be delivered to the areas in need because the special FO membranes cannot be created out of any regular material. Unlike water bottles, however, these packs are much lighter and take up much less space (HTI, 2010).

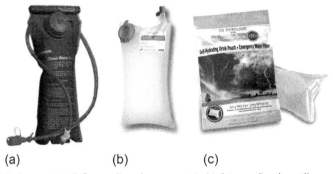

(a)   (b)   (c)

Figure 2.5 Symmetric (left panel) and asymmetric (right panel) solar stills used to distil drinking water from sea water or other low-quality water. *After Saidur et al. (2011), original concept from Al-Hayek and Badran (2004).* (For the color version of this figure, the reader is referred to the online version of this chapter.)

2.3.3 Commercial Products

Two commercial manufacturers of FO membranes are Hydration Technologies Inc. (HTI) and Catalyx Inc. Most academic research studies on FO have obtained their membranes from HTI. The manufactured membranes' properties have been tested by independent research companies. While research was conducted on the membranes created by the two manufacturing companies, no independent research has been conducted on any of HTI's proprietary emergency water purifiers.

2.3.4 Costs

- Expedition = $300 for filter, bladder, and 10 syrup pouches
- Hyrdopack = $2–4.50 for individual packets when bought in bulk
- X-Pack = $57.95 for pouch and syrup pouches, $2.30 for individual syrup refills

2.3.5 Evaluation

For first responders to disasters or for short periods, a pack using forward osmosis is a safe and useful technology that provides clean water as well as nutrients to those in need. It is more cost effective and sustainable than bottled water but is not applicable for long-term use and cannot provide water for cooking and washing.

CHAPTER 3

Solar Pasteurization

Contents

3.1	Microbiology of Water Pasteurization	31
3.2	Use of Solar Cookers for Drinking Water Production	33
3.3	Devices Designed Specifically for Water	36
3.4	Simple Devices from Common Materials	38
3.5	Commercial Devices in Production	41
3.6	Devices with Recovery Heat Exchange	41
3.7	Water Pasteurization Indicators	44
3.8	Multi-use Systems	47
3.9	The Greenhouse Effect	47
3.10	Use of SOPAS in Conjunction with SODIS	49
3.11	SODIS and Titanium Dioxide	49
3.12	SOPAS and SODIS Technology Evaluation	50

Keywords: Water pasteurization, Solar cooker, Heat exchange, Greenhouse effect, Solar disinfection, Solar pasteurization

3.1 MICROBIOLOGY OF WATER PASTEURIZATION

Heating water to a sufficiently high temperature for a certain period destroys harmful microorganisms; this process is often termed "pasteurization" after the scientist Louis Pasteur (1822-1895). The required temperatures and corresponding periods for the destruction of microbes are shown in Table 3.1 (Ciochetti and Metcalf, 1984). The D value is the time to kill 90% of the organisms. A time of $5D$ would kill 99.999% of the microbes.

At temperatures higher than those listed in Table 3.1, the D value decreases significantly; for example, for the microbes in Table 3.1 with a D-value of $\sim$1 min at 60 °C, the D-value decreases to $\sim$12 s at 65 °C. Also, considering that water as it heats toward 65 °C also spends time at 64 °C, 63 °C, and so on (at which temperatures additional fractions of microbes are killed), it is widely agreed that pasteurizing water to 65 °C will make it safe to drink. Many pasteurization devices use a safety factor of several degrees above 65 °C to allow for variations in measurement and equipment.

☆"To view the full reference list for the book, click here"

†Adapted from Pejack (2011).

Table 3.1 Effect of Temperature on Microbes (Ciochetti and Metcalf, 1984)

Temperature	Microbes Destroyed	Time Required for Inactivation	Total Time for Pasteurization (h)
55 °C (131 °F)	Worms, protozoa	1 min	4.1
60 °C (140 °F)	E. coli, rotavirus, Salmonella typhi, Vibrio cholera, Shigella	1 min	4
65 °C (149 °F)	Hepatitis A virus	12 s	4.5

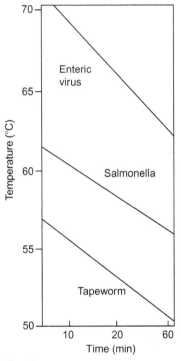

Figure 3.1 The semi-log plot of time versus temperature required to pasteurize water (redrawn based on Feachem et al., 1983).

The time required to pasteurize water decreases exponentially with an increase in temperature (Feachem et al., 1983), as illustrated in the semi-log plot of time versus temperature shown in Figure 3.1.

It is a common misconception that water pasteurization requires boiling water (sometimes it is even stated that the water must boil for 20 min). Unnecessary boiling wastes a significant amount of fuel, which is already an expensive and scarce item in much of the world. Boiling water requires about twice the energy as heating it to 65 °C, plus the extra time for monitoring it and the time and expense of obtaining fuel. The idea that water must be boiled to be pasteurized may have arisen because it is easy to tell when water boils, and one can then be certain that the water has been heated above 65 °C.

Note that the type of pasteurization discussed in this chapter is concerned with the destruction of microbes only. Toxic chemicals, metal salts, and other contaminants are not removed by pasteurization. Insoluble material such as silt can be decreased by prefiltering or settling the water.

Pasteurization as discussed in this chapter is a thermal process, relying on the destruction of microbes by temperature (achieved with solar energy). A related but different water treatment concept is solar disinfection (acronym SODIS), where microbes are destroyed by the direct action of certain wavelengths of the solar spectrum, independent of temperature. It is the ultraviolet range of the spectrum (200-400 nm) that is most effective in destroying microbes; consequently, transparent containers that have good transmittance in this wavelength band should be used. The interested reader can find many reports of SODIS studies and trials, for example, Caslake et al. (2004) and Meierhoffer and Wegelin (2002).

Methods of solar pasteurization (SOPAS) discussed in this chapter include:
1. Solar cookers
2. Bottles and reflectors
3. Commercially produced devices
4. Flowthrough devices
5. Devices that use the greenhouse effect
6. SOPAS used in conjunction with SODIS
7. SOPAS used in conjunction with titanium dioxide

3.2 USE OF SOLAR COOKERS FOR DRINKING WATER PRODUCTION

There is a widely disseminated body of knowledge and literature on the subject of solar cooking and solar cookers, and several international conferences have been held on this subject in the past 20 years; proceedings of these conferences have been archived by Solar Cookers, International (Sacramento, CA). Because many foods are cooked in the range of 80-100 °C, it is not surprising that solar cookers are also used to pasteurize water at 65 °C.

Perhaps the first type of practical solar cooker, the "box cooker" is commonly used around the globe. In its simplest form, it is a box insulated on its sides and bottom that has a transparent top cover. The box also frequently

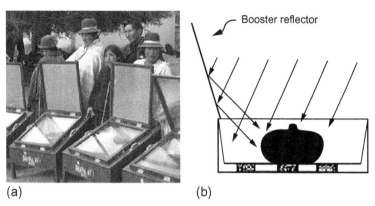

(a)   (b)

Figure 3.2 The box cooker. (a) Wooden box cookers with slanted covers, made in Bolivia. (b) Cross-sectional schematic drawing of a box cooker. *(Ray and Jain, 2011).*

has a "booster reflector" at the top to augment the solar power entering the top cover, as shown in Figure 3.2a.

In recent decades, numerous analyses of box cookers have been made, and clever design attributes have been developed, which will not be discussed here. Some of the relevant design and operating parameters are:

- Thermal insulation of box
- Optical properties of top cover
- Surface characteristic of box interior
- Booster reflectors
- Solar power, cloudiness
- Time (hour, day, month)
- Latitude
- Device orientation, tracking/nontracking
- Mass and type of food
- Pot size, surface, lid
- Wind, ambient temperature

Various box cookers are in use around the world; some are small and hold one pot, others hold more than 10 pots. Box cookers (Figure 3.2) have been made from cardboard (one of the first was the famous Kerr-Cole cooker), wood, metal, plastic, and mud. Thermal insulation for the box is necessary, and, again, very many materials have been used, including crumpled paper, air space, paper baffles, feathers, rice hulls, and foam.

The simplest way to use a box cooker to pasteurize water is to place a pot of water in the cooker as if cooking food. Often, a reasonably designed box (able to cook a variety of foods in 2 h) with a solar radiation of approximately 700 W/m^2 can pasteurize a liter of water in about an hour.

Another popular cooker design, the "panel cooker," consists of several flat reflective surfaces (panels) adjacent to a pot, which is enclosed in a transparent bag or enclosure. There is no insulated box; rather, the heat loss from the pot is minimized by the layer of hot air between the pot and the bag. Bags are commonly made of polypropylene film, but other transparent films have been used (such as polyethylene and nylon). The bag should not fit tightly on the pot; this prevents the bag from melting from the hot pot and prevents too much heat loss. If the bag has too large an air space, the insulating ability of the air space is decreased. An air gap of about 1-2 cm is recommended. Ideally, the bag's material would have a high transmittance for solar radiation in wavelengths of 0.3-1 nm and low transmittance for thermal radiation in wavelengths of 5-14 nm (around 100 °C). Two types of panel cookers are shown in Figure 3.3.

The panels are reflective surfaces that vary in number and orientation. The panels' function is to reflect solar rays onto the bag, increasing the concentration and solar power to the bag (and the water pot). The number of panels ranges from 1 to a 12 or more, and the panels are usually flat planes for ease of construction and foldability; however, curved, hemispherical, and other modified panels have been employed.

A particular version of the panel cooker, the "CooKit" (Figure 3.3), developed by Solar Cookers International (SCI, 2009a,b), has the desirable attributes of low cost and weight, ease of construction, and foldability. It has seven flat panels when unfolded, and the front flap is adjusted upward to reflect sunlight on the pot or bag. One useful characteristic of this device is that it does not need to be re-aimed at the sun for several hours because the panels

(a)   (b)

Figure 3.3 (a) The CooKit (SCI, 2009a,b) with a pot inside a plastic bag. (b) The "Hot Pot" has a glass bowl instead of a bag. There is an air space between the inner black pot and the bowl. *Panel (b) photo credit: Patricia McArdle.*

compensate for the relative solar motion. A pot can be filled with polluted water and, with sufficient solar radiation and time, the CooKit can pasteurize the water in approximately the same amount of time as the box cooker.

A large variety of panel cookers have been developed and produced in recent years. Design variations include the number and shape of the panels, overall size, type of container, materials used in construction, and method of folding/unfolding.

A third class of solar cooker, called a "concentrator," concentrates the solar radiation much more than a box or panel cooker. This type uses parabolic surfaces (simple or compound), Fresnel lenses, or multiple aimed reflectors. Heat rates are much higher for concentrators than for box or panel cookers. Thus, users must be careful not to place their arms or face into the concentrated solar rays. Because the radiation to the pot is so high, a bag or cover around the pot is usually unnecessary. To use the reflective properties of parabolic surfaces most effectively, concentrators should be tracked to the sun more often than box or panel cookers. Examples of concentrators are shown in Figure 3.4.

Figure 3.4 Example of a concentrator. A pot of water is placed on a ledge in the center of the parabolic surfaces to take full advantage of the sun's rays. *(Ray and Jain, 2011).*

3.3 DEVICES DESIGNED SPECIFICALLY FOR WATER

While devices that are used for solar cooking can be used to pasteurize water, devices that are designed specifically for water treatment can be more efficient. These devices incorporate factors that minimize the time required to pasteurize water in their design.

The time required to pasteurize water is decreased by:

- Increasing solar input power by operating when the solar power is highest (e.g., around solar noon or at a sunny location) or by using reflectors to achieve concentration.
- Decreasing the mass of water in the device. If the initial mass is too great, the pasteurization temperature may not be achieved before the solar flux becomes too low.
- Decreasing the initial temperature difference between the pasteurization temperature (e.g., 65 °C) and the initial temperature of the polluted water. Starting with warmer water decreases the time required to pasteurize it.
- Minimizing heat loss from the water as it heats by using thermal insulation, paying attention to thermal radiation from the device, and avoiding windy locations.

This simple model of a pasteurizing device ignores the effect of water evaporation. If water vapor is allowed to escape the device, this can represent a huge energy loss, increasing the time to pasteurize the water. As noted above, minimizing the heat loss decreases the time required to pasteurize water. A very low loss factor can be achieved by vacuum insulation, which practically eliminates heat loss by convection. One of several such systems (Figure 3.5) is the Solar Kettle (Kee, 2006), where the water is contained

Figure 3.5 An array of Solar Kettles articulated to pour out the pasteurized water. *Photo credit: Alex Kee.*

in an insulated glass vacuum tube 0.75 m in length. The outer surface of the inner tube is coated with a selective material (such as aluminum nitrate) to create a selective surface (as shown in Figure 3.5), enhancing the absorbance of solar rays and minimizing re-radiation of thermal radiation.

3.4 SIMPLE DEVICES FROM COMMON MATERIALS

Surprisingly simple devices made from common materials can pasteurize small quantities of water in 1 h. Here we discuss simple, common, inexpensive pasteurization devices because, often, poor, distressed peoples do not have safe water nor do they have the resources to employ sophisticated, complex water treatment devices. Perhaps the simplest pasteurization device is a simple flat puddle of water with a transparent cover that is exposed to sunlight, with or without adjacent planar reflectors (Andreatta, 2009; Pejack et al., 1996).

Researchers obtained experimental results using this simple flat type of puddle; a transparent polypropylene bag to contain the water; and zero, one, or two adjacent vertical reflectors each made of 0.30 m × 0.30 m (1 ft × 1 ft) of aluminum foil. When two reflectors were used, they were oriented 90° to each other with the water between the reflectors. The size of the flat water puddle was 0.25 m × 0.41 m (10 in. × 16 in.), and the puddle rested on a foam surface to minimize heat loss from the bottom. The experiments were performed in June in Stockton, California, on a sunny day when the noon solar flux was 980 W/m^2.

Andreatta (2009) describes an amazingly simple method to pasteurize the flat puddle, called the "sack pasteurizer." Starting with a circular sheet of transparent plastic about 1 m in diameter, the plastic is bunched up to form a sack that holds 3 l of water and then tied closed. The sack is placed on various surfaces (black plastic on grass, foam, or bubble wrap, grass alone, black foam), which shapes the water into a somewhat flat puddle, and the sack is then covered with a separate sheet of transparent plastic similar to that which forms the sack. An air gap is left between the sack and the cover. Experiments at 40°N latitude (Ohio) in late August on a strong sunny day achieved pasteurization temperatures (Table 3.2).

Another simple water pasteurizer can be made with easily obtained cylindrical bottles. The configuration consists of two vertical reflective surfaces of aluminum foil oriented 90° to each other, plus a third reflector on the ground (Figures 3.6 and 3.7). The total reflector area was 0.36 m^2 (3.9 ft^2).

Table 3.2 The Test Results of a Puddle Pasteurizer

Time to Pasteurize (min)	Water Depth (cm)	Water Volume (l)	Reflectors
25	0.40	0.41	Two
79	0.98	1.00	Two
54	0.98	1.00	One
109	0.98	1.00	Zero

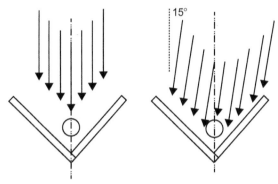

Figure 3.6 Bottle pasteurizer with 90° vertical reflector planes. At left, solar rays are on centerline, at right, they are at 15° azimuth. *(Ray and Jain, 2011)*.

Figure 3.7 A 1-l PET bottle inside a cover made from two 2-l PET bottles (Bottle 2, Cover 2). The reflector in front is tilted up. *Photo credit: Ed Pejack.*

The water containers used were:

- Bottle 1: 0.36 l glass, spray painted black
- Bottle 2: 1.0 l soda (polyethylene terephthalate, PET), painted black
- Bottle 3: 0.5 l brown glass beer bottle

Transparent covers for the bottles were made in these ways:

- Cover 1: Polypropylene bag
- Cover 2: Two 2-l PET bottles slit around the middle, then one bottom half slit
    6 × lengthwise and the other bottom half slid over the other
- Cover 3: No cover

Results of the experiments (on the same day as the flat puddle experiment discussed above) were:

Time to Pasteurize (min)	Container
116	Bottle 1, Cover 3
82	Bottle 2, Cover 2
72	Bottle 2, Cover 1
51	Bottle 3, Cover 1

Sometimes a question is raised concerning whether exposing PET plastic bottles to sunlight could cause plasticizers to enter the water. Studies have indicated that SODIS devices using PET are safe with respect to human exposure to the plasticizers di(2-ethylhexyl)adipate (DEHA) and di(2-ethylhexyl)phthalate (DEHP) (Yegian and Andreatta, 1996).

Reflectors can also be used in conjunction with PET bottles. Research found that two 90° reflector planes could be oriented with an azimuth of 15°; then, for 2 h, the relative solar movement would result in the concentration decreasing from 3.6 to 3.0 for the first hour, then increasing from 3.0 back to 3.6 for the second hour, without any adjustment of the reflectors. In this case, the water cylinder was placed two bottle diameters from the intersection of the two reflectors. These studies suggest that 0.5-1 l of water may be pasteurized in 1-1.5 h using simple materials and simple geometry. Too often, large numbers of people such as refugees and victims of catastrophes are congested in a confined area with contaminated drinking water. In those situations, methods similar to those described above may be very advantageous. Travelers in remote regions where safe drinking water is unobtainable may also find these methods very useful.

Many variations of simple pasteurization devices have been developed using various configurations of water containers, covers, and reflectors (Cengel, 1998; Duff and Hodgson, 1999; Yegian and Andreatta, 1996).

3.5 COMMERCIAL DEVICES IN PRODUCTION

A form of the puddle-type pasteurizer described above is commercially produced as AquaPak® (Figure 3.8a), which is a small (0.35 m × 0.35 m), flat polyethylene plastic bag with one side made of transparent bubble pack insulation, and the other side black. Its ability to pasteurize water depends on the ambient temperature, solar power, and the quantity of water to be pasteurized (2-5 l). It is reported to have been used in 30 countries. When the solar radiation was 800 W/m² and the initial water temperature was 25 °C, 2 l of water were pasteurized in 2.2 h. The lid of the AquaPak has a small, wax-filled capsule that indicates (by displacement of the melted wax) whether the water has been pasteurized. A larger capacity water pasteurization system, the SunRay 30 (Figure 3.8b), is a flat collector (79 cm × 61 cm × 9 cm) that contains 10 black bottles. It pasteurized 7.5 l of water in 1.5 h.

3.6 DEVICES WITH RECOVERY HEAT EXCHANGE

In the devices discussed above, pasteurized water is either removed at the pasteurization temperature or allowed to cool in the device for use later. The energy in the pasteurized water, once used to heat the water to pasteurization temperature, can be transferred to water not yet pasteurized, thereby requiring less solar energy to heat more water. Such a device uses a heat exchanger, which makes the device a flowthrough device, as opposed to a batch-process device. The savings in energy or the increase in quantity of pasteurized water can be quite significant.

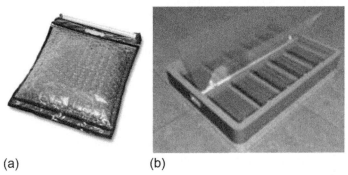

(a)   (b)

Figure 3.8 Commercialized pasteurizers. (a) AquaPak and (b) SunRay 30. *Photo credit: left: Solar Solutions; right: Safe Water Systems.*

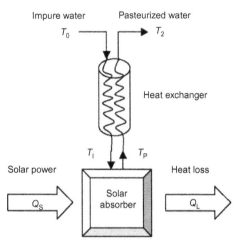

Figure 3.9 Essential features of a flowthrough pasteurizer.

The flowthrough pasteurizer works as follows: As shown in Figure 3.9, polluted water at ambient temperature T_0 enters one side of the heat exchanger at the mass flow rate $\dot{m}$ and exits the heat exchanger at the higher temperature T_1, where it enters the solar device. In the solar device, solar power Q_s enters and some power Q_L is lost to the ambient environment. The flowing water achieves pasteurization temperature T_p in the device, exits the device, and enters the other leg of the heat exchanger. The water flows counter-currently, transferring heat to the incoming stream, and finally exiting the heat exchanger at temperature T_2, which is lower than T_p but slightly higher than T_0.

A reader interested in heat exchanger theory, design, and analysis may consult any of the many references on the subject (Duff and Hodgson, 2005). Other analyses of pasteurizers with heat exchangers can be found in Stevens et al. (1998) and Duff and Hodgson (2005).

Whereas the batch-process pasteurizers function with a fixed mass of water, the mass flow rate in the flowthrough process can be adjusted during operation to account for the variable of solar power. The batch process in most cases uses a fixed mass of water, and, if set at too high a value or if the solar power decreases during heating, the entire batch may not be pasteurized. Similarly, if the solar power is higher than expected, the mass of water pasteurized could be less than is possible with a larger starting mass.

The flowthrough pasteurizer often uses a thermostatic valve, which opens and closes in response to temperature, to control the water flow rate. Some applications have adapted the thermostatic valve from the cooling systems of automobiles, with the notion that the valve would allow flow only when the temperature is above the pasteurization temperature.

The valves operate from closed to open over a temperature range, making the mass flow rate vary with temperature, opening more at the higher temperatures and closing more at the lower end of the range, sometimes with cyclic operation. This cyclic operation can lead to inefficiencies in the device as the valve is not able to successfully control the flow of the heat exchanger throughout the entire day. Changes in ambient temperature cause the valve to increase flow before water has been pasteurized or decrease the flow causing boiling and contamination issues. (Duff and Hodgson, 2005). Issues with thermostatic valves have led to research on more robust techniques for flow-through solar pasteurization.

Duff and Hodgson (1999, 2005) have designed and operated several versions of a heat exchanger pasteurization system in which the flow is controlled by the density difference between hot (pasteurized) water and colder (unpasteurized) water, rather than using valves. Heated water is directed to a vertical riser tube, and, when the water there is hot enough, the lower density (and higher water column) allows the pasteurized water to "spill over" and flow through the heat exchanger. The system eliminates the operation problems of thermostatic valves. Some of the performance data reported includes an average hourly production rate of 16.4 kg/h (1 kg water = 1 l water) for a 2-h period when the normal solar radiation averaged 913 W/m^2 on a mostly sunny day. The production of pasteurized water for that entire day, from 8:00 am to 3:00 pm, was 86 l. Duff and Hodgson's collector was a set of five evacuated heat pipes with a total area of 0.45 m^2, which had a relatively low heat loss coefficient.

Many heat exchangers are the concentric tube or flat type (Figure 3.10). Yegian and Andreatta (1996) discuss their experiences with a concentric

Figure 3.10 A commercial pasteurizer, the SunRay 1000. *(http://www.safewatersystems. com/sunray_1000).* (For the color version of this figure, the reader is referred to the online version of this chapter.)

tube heat exchanger consisting of an inner and an outer annulus separated by a copper tube. The outer annulus contains wrapped wire to enhance the heat transfer. These authors also discuss a flat heat exchanger made with sandwich-like layers of copper sheets and wood. Rubber strips force the water to take a serpentine path to increase heat transfer. Heat transfer effectiveness values for the two designs were 60-80%, depending on the water mass flow rate. The mass of water pasteurized using these devices compared to the batch-process-type device was increased by a factor of eight. A system can be assembled using a solar collector originally designed for service water (e.g., for heating buildings) and adapting a counterflow heat exchanger. However, there are several disadvantages to using a flowthrough pasteurization device. For example, the thermostatic valve can break.

3.7 WATER PASTEURIZATION INDICATORS

A useful attribute of many batch-process solar water pasteurization devices is that once they are positioned in the sun, they may be left unattended. However, suppose that a device is left alone in strong sun, and the user returns late in the day. Did the water, now cool, reach pasteurization temperature during the day? Unless a recording thermometer has been installed, the temperature history of the water is unknown. What is needed is a sensor that can indicate, and preserve that indication, whether the pasteurization temperature was achieved during the day. Such an indicator could be based on a number of physical phenomena, such as:

1. Melting of a material from solid to liquid: A material that melts at the pasteurization temperature could be arranged so that once melted, it changes shape or location and maintains that change when the temperature later decreases.

2. Differential thermal expansion: Two materials with different coefficients of thermal expansion could be made to interact so that a detectable change in geometry is produced at the pasteurization temperature, and that change is maintained when the temperature later decreases.

3. Color change: Certain materials change color at specific temperatures. If the material changes color at a certain temperature, that could indicate the pasteurization temperature has been reached.

One must be aware that, in a volume of water, there will likely be temperature variations. Temperatures at the bottom of a container may be several degrees colder than at the top. Therefore, the temperature indicator should be placed at a lower level of the water. In cases where the container is heated from the bottom (as for example, in a concentrator), the spatial temperature

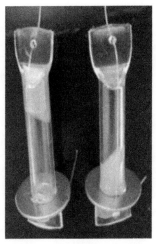

Figure 3.11 Two views of a WAPI. The image on the left shows the wax at top. After reaching melting temperature, the wax migrates to the bottom, as shown on the right. The metal washer slides on the tube to keep the tube vertically suspended by the string. *(Ray and Jain, 2011)*. (For the color version of this figure, the reader is referred to the online version of this chapter.)

variation may be quite different. In any particular situation, spatial temperature variations should be measured, and the indicators should be placed at the colder regions.

A device based on a material that melts from a solid to a liquid has been widely used in a variety of SOPAS devices, in several forms. The WAPI (water pasteurization indicator) (SCI, 2009a,b) is comprised of a small quantity of wax (soybean wax, for example, which melts at ∼70 °C) inside a plastic (or glass) tube sealed at both ends, as shown in Figure 3.11. Initially, the wax is at one end of the tube. If the WAPI is suspended vertically with the wax end on top, then as the melting temperature of the wax is reached, the wax melts and flows by gravity to the lower end of the tube. Thus, finding the wax at the bottom indicates that the water has been pasteurized, even after the water has cooled. The WAPI can easily be reset for further use by inverting it. It has a flexible string and a weight so it can be suspended vertically in water. The WAPI is currently used in many locations worldwide.

The water temperature indicator shown in Figure 3.12 is based on differential thermal expansion of dissimilar metals and is called the Snap Disk (Saye and Pejack, 1994). It is constructed of two dissimilar stainless steel disks that are pressed together and bonded to form a 2.54-cm disk. At any temperature, the disk is concave on one side and convex on the other. As the temperature is raised and reaches the snap temperature, the disk snaps into

Figure 3.12 Two snap disks with folded housing ($\sim$2.5 cm^2). *Photo credit: Ed Pejack.* (For the color version of this figure, the reader is referred to the online version of this chapter.)

reverse concavity (an example of what is called "snap buckling" in mechanics of materials). The disk remains in the new snapped position even when the temperature decreases below the snap temperature. The snap temperature is quite precise and repeatable over millions of cycles. Snap disks are manufactured in many countries. Because this is a relatively mature technology, disks can be made with any snap temperature and tolerance.

To use the snap disk, one snaps the disk to the starting position and places it in the water container. When the water is heated to the snap temperature, the disk automatically snaps to the opposite side. Then, at any time later, when the water has been cooled, the user can see if the disk has snapped and therefore tell if the water was pasteurized because the disk does not snap back when cooled. To reuse the disk, users push on the disk with their fingers to snap it back to the starting condition. The disk is fast acting, and, if lowered into water hotter than the snap temperature, it immediately snaps with a distinct audible "pop."

Of course, the user must be able to distinguish between the two sides of the disk. In the Snap Disk shown, the disk was placed in a folded stainless steel housing that is open on one side and has a hole on the other side for manually snapping the disk, making the two sides of the disk readily discernible. The housing also prevents users from overstressing the disk when they manually reset it. The disk has a snap temperature of 68 °C with a manufacturer's tolerance of 2.8 °C, but tests on samples showed the tolerance to be much less. In actual use, after a time the Snap Disk would experience some corrosion from being exposed to the water chemistry as well as the intimate contact of dissimilar metals. Therefore, the disk should be dried when not in use. The manufacturer also reports that other metal combinations could provide better corrosion resistance.

3.8 MULTI-USE SYSTEMS

In addition to the solar water pasteurization applications (~65 °C) described above, other uses for solar thermal devices include:

- Autoclaving for medical sterilization (>122 °C)
- Producing distilled water for medical use
- Producing service hot water (50 °C)
- High temperature steaming and cooking (>150 °C)

It is possible to create a multi-use system designed for several of the purposes listed above. Such a multi-use system, using solar collectors, heat exchangers, and associated valves and piping, conceivably could be more practical to design and construct than individual systems for all the uses given above. The higher temperatures needed would likely require low loss collectors (e.g., vacuum tube collectors) and solar concentrators, and the design and operation would be considerably more complex and expensive than simple pasteurization devices.

3.9 THE GREENHOUSE EFFECT

Several SOPAS devices use the greenhouse gas effect (GHE) to create a more efficient heating process. A typical example of a SOPAS device is a water bottle filled with contaminated water and left in the sun for several hours. Sealing the water bottle inside a clear plastic bag uses the greenhouse effect to make the SOPAS process more efficient. Several authors have researched GHE devices that consist of lids made out of glass and sides made out of reflective materials to create a chamber to hold clear water bottles (Jagadeesh, 2006; Valenzuela et al., 2010). The glass lid traps heat inside the chamber, keeping the water bottles warm and protecting them from wind. See Figures 3.13 and 3.14 for examples of GHE devices. An advantage to using the GHE is that this technology can operate with lower intensity radiation, that is, on cloudy or winter days. PET bottles are generally used to pasteurize water because they are easily accessible in many rural areas, but empty wine bottles have also been used (Jagadeesh, 2006). While health concerns have been raised regarding the leaching of toxins into water from heating PET bottles using sunlight, Schmid et al. (2008) notes the health risk does not reach critical levels. The internal water temperature needs to exceed 66 °C for a certain period for inactivation of Hepatitis A (Kang et al., 2006). When this temperature is met, other bacteria and pathogens

Figure 3.13 Example of a greenhouse effect device (Valenzuela et al., 2010).

Figure 3.14 Example of a greenhouse effect device using wine bottles (Jagadeesh, 2006). (For the color version of this figure, the reader is referred to the online version of this chapter.)

will also be inactivated. There are discrepancies in the research about how long inactivation temperatures must be sustained, varying from 1 min to more than 1 h (Jagadeesh, 2006; Kang et al., 2006; Valenzuela et al., 2010). Maintenance for this technology is minimal and includes cleaning the PET bottles or replacing the bottles when they become brown or scratched (Jagadeesh, 2006). GHE devices operate similarly to solar cookers although they are designed specifically to pasteurize water in bottles. Just as with solar cookers, a GHE device does not require skilled operators though the bottles require careful handling because improper water storage could lead to the growth of microbes (Kang et al., 2006). Costs of these devices were not discussed although all materials were noted to be accessible in developing countries and easily assembled.

3.10 USE OF SOPAS IN CONJUNCTION WITH SODIS

SODIS uses the sun's radiation (UV-A rays) rather than heat to inactivate microorganisms in water. The thermal and irradiation processes can be used in conjunction to make water treatment quicker and more efficient. SODIS is dependent on the intensity of solar radiation, a large surface area exposure, and shallow water depths. The effectiveness of SODIS is also dependent on low turbidity because particulates in water cause the sun's rays to scatter. Using both SOPAS and SODIS creates a synergistic effect. Once the temperature 45 °C is surpassed, inactivation of microorganisms starts to occur; in SOPAS, temperatures need to be in the high 50 or 60 °C range. Maintaining a water temperature of approximately 50 °C requires only one-third of the typical solar irradiation intensity needed for inactivation (Oates et al., 2003). Because SODIS is dependent on the intensity of solar radiation, the material of the container used to purify the water is important. Some polymers can block UV-A rays, so the material needs to be a UV-transparent, colorless plastic or glass. Several small-scale technologies have been studied for using SODIS/SOPAS in conjunction, including water bottles (Hindiyeh and Ali, 2010), solar cookers (Saitoh and El-Ghetany, 2002), and specially adapted flowthrough devices (Caslake et al., 2004). A disadvantage to this procedure is that no indicator exists to show when disinfection has been achieved. The current indicator of disinfection is given as a length of time the water needs to irradiate depending on latitude and solar conditions. These general guidelines may not always be applicable to every situation. While SOPAS has various indicators for when the critical temperature is reached (including melting wax, shown in Figure 3.11, and popping metal, shown in Figure 3.12), no indicators have been developed to distinguish disinfected water from untreated water.

3.11 SODIS AND TITANIUM DIOXIDE

The efficiency of SODIS can be improved through use of titanium dioxide (TiO_2). TiO_2 has been found effective in water purification because TiO_2 works as an advanced oxidizer that operates in conjunction with the sun's UV-A rays. A PET water bottle is suggested as a container. A TiO_2 film can be created by acidifying TiO_2 with $HClO_4$; then the inside of the bottle can be coated with the TiO_2 solution or the solution can be applied to glass rings that are then suspended in the bottle. In developing countries, coating the inside of the bottle is easier than coating glass rings. The TiO_2 film makes the irradiation process more efficient when the water temperature is high and when there are high quantities of dissolved oxygen (DO) in the water. High temperatures can be achieved by placing the water container in direct sunlight and adding reflective surfaces. Shaking the PET bottle vigorously when three-quarters full and then completely filling the bottle can create the necessary high DO values. Low-turbidity water provides a clear medium for penetration of UV-A rays, so high-turbidity water should be filtered. The time recommended for sun exposure varies. Exposure time should be approximately 2 h when starting radiation in the morning on a sunny day. On cloudy days, exposure needs to last around 7 h (Heredia and Duffy, 2006). These times decrease when reflectors are added to the system (Gelover et al., 2006). While the length of needed sun exposure is still long, the TiO_2 does decrease the time needed for inactivation compared to simple SODIS. Several advantages to TiO_2 films exist apart from increasing the rate of inactivation. Adding a TiO_2 film can oxidize other chemicals in the drinking water, such as arsenic. Another advantage of using TiO_2 films is that there is no regrowth of bacteria after treatment and during storage. Although TiO_2 can have advantages, several improvements still need to be made to the process. Acidifying the TiO_2 can be dangerous for those who are not used to handling strong acids. Heredia and Duffy (2006) suggested that a medical technician in the town could be the one to handle the acid; however, future advances are needed to make this technology more manageable. A disadvantage of using a coating in a bottle is that the reaction rate of oxidation is limited because only the water close to the edge of the bottle interacts with the thin coating. Similar to SODIS, this method also has no indicator to tell the user when the water has been disinfected; such an indicator needs to be developed. No cost estimates were found for the use of TiO_2, and there was no mention of how much powder would be needed for a community's water treatment processes.

3.12 SOPAS AND SODIS TECHNOLOGY EVALUATION

A SOPAS system can have many forms, as seen from the above discussion, and some thought, analysis, and planning is required to decide what is appropriate, practical, and feasible for any particular situation.

When choosing a solar water pasteurization system, keep in mind that there is a trade-off between using commercially available devices and adapting raw materials to construct custom devices using the basic concepts outlined in this chapter. Commercial SOPAS devices have desirable attributes in that they are more likely perform as intended, have a history of operation, have a robust design, have clear operation procedures, and reduce the time needed for disinfection. However, because of variable local conditions, even well-developed commercial devices need to have some support from the manufacturers. Commercial devices may not be feasible in some cases because of purchase cost, transportation, and availability of spare parts. In those cases, a custom design and unique construction may serve the users better and may be more environmentally appropriate.

An example of a SOPAS system for sustainable drinking water is the Safe Water Project (Kenya, 2008), which was initiated by Solar Cookers, International (Sacramento, CA), and directed by Dr. R. Metcalf (SCI, 2009a,b). Among the important factors to consider are the end users, who can be:

- One or two persons (at home, hiking, or trekking)
- One family
- An extended family
- A village
- A commercial enterprise

SODIS is currently used by 4.5 million people in developing countries, the majority of those very low income (McGuigan et al., 2012). Although SODIS is recommended by the World Health Organization as a short-term emergency water treatment (McGuiellen et al., 2012), very few studies have actually assessed the use of SODIS during an acute emergency (Lantagne and Clasen, 2009). Alternatively, McGuiellen et al. (2012) argues that the effective use of SODIS requires behavioral changes that are only effective in the acute phase of an emergency. These behavioral changes include cleaning, filling, organizing, and collecting the SODIS water that must be drunk consistently to decrease diarrhea. It is easier to encourage behavioral changes during an emergency, but these changes are typically short lived and are not always sustainable in the long term. For areas that are already familiar with SODIS, it might be possible to effectively implement the technique in the short term and the long term.

Someone implementing a solar water pasteurization system needs to consider factors relating to the acute emergency phase response as well as continued, sustainable operation that is based on the needs of the area. For example, the availability of materials to construct additional devices or to repair and replace parts should be included in the evaluation. Finally, the end users need to have some training to ensure successful operation, repair, and evaluation of the system.

A summary of the technologies discussed above can be found in Table 3.3.

Table 3.3 Summary of Simple Solar Technologies

	Base Technology	Improvements to Technology	Time of Year Tested	Time of Day/Solar Intensity	Time to Reach Inactivation	How Inactivation Is Quantified	References
Extra reflective surface	SODIS: 1.5 l PET bottle	Black paint applied to elongated bottom half, shaken, set on rooftop	Haiti: January	Average peak intensity of 651 W/m²	1-day exposure provided 50% removal, 2-day exposure provided 100%	Remove 100% of indicated organisms	Oates et al. (2003)
	Solar pasteurization: solar box cooker	3.7 l jugs painted black, cooker repositioned to the sun every so often	Sacramento, CA: sunny days from March–August	10 am–5 pm	4–5.5 h	No info	Ciochetti and Metcalf (1984)
	Puddle pasteurizer	AquaPak, SunRay 30		800 W/m²	2 l in 2.5 h, 7.5 l in 1.5 h	99.99% removal of pathogens	Pejack (2011)
	Solar pasteurization: 1.5 l PET bottles	Mirrors	Jordan: June–November	10 am–6 pm 167-361 J/cm²	4 h	>1.1 MPN/100 ml	Hindiyeh and Ali (2010)
Greenhouse effect	SODIS: wine bottles shaken for aeration	Greenhouse effect: bottles inside a reflective box w/clear cover	India: no info	10 am–3 pm	5 h	4-log 10 reduction of E. coli	Jagadeesh (2006)
	Solar cookers	Box, panel, or concentrator cookers		700 W/m²	~1 h		Pejack (2011)
	Vacuum tube insulator	Solar Kettle	New Zealand: November	Ambient Temp = 18 °C	~2–3 h	No info	Kee (2006)
	SODIS: glass bottles shaken for aeration	Greenhouse effect: bottles inside a reflective box w/clear cover	Mexico: Jan 27-31	8 am–6 pm	5 h	Presumptive and confirmatory results negative for coliform	Valenzuela et al. (2010)

TiO_2 films	SODIS: 2l PET bottles in reflectors w/ aluminum foil	TiO_2 coating on glass cylinders dangled in PET bottles	January–July	Solar noon, 800 W/m²	15-20 min, less than 2 h	>1 MPN/100 mL	Gelover et al. (2006)
	SODIS	TiO_2 coating to inside of PET water bottle	Michigan: September–October	12-4 pm	10 am-12 pm (2 h) or 12 pm-4 pm (4 h)	1-160 kJ/m²	Heredia and Duffy (2006)
Flowthrough devices	Solar pasteurization: flowthrough device based on density principles	Doesn't use thermostatic valves	Fort Collins, CO: March	7 am-3 pm, cloudy (64-140 W/m²) and sunny days (195-1008 W/m²)	Cloudy: from 7 am to 3 pm can produce 35 kg/day; sunny: from 7 am to 4 pm can produce 86 kg/day	No bacteria testing considered	Duff and Hodgson (2005)
	Solar pasteurization flowthrough device	Commercialized in India	Velore, India: February–August	No sun radiation measurements	Sunny days: 2 h; cloudy days: 4 h	No info	Kang et al. (2006)
	Flowthrough device w/heat actuated valve	SunRay 1000	No info	No info	Sunny days: 9 am-4 pm, produces avg. of 264 gallons	99.999% of microorganisms	Safe Water Systems (2012)
	SODIS: flowthrough device w/shallow snake groves	Conjunction w/SOPAS	Pennsylvania: June, July, August	11 am-3 pm, Temp=22.9-33°C, rad=500-800 W/m²	20-60 min	<1 Coliform/100 ml of water, 4 log10 U reduction in bacterial load	Caslake et al. (2004)

CHAPTER 4

Disinfection Systems

Contents

4.1 UV Light Systems — 55
  4.1.1 Advantages and Disadvantages — 71
  4.1.2 Design Considerations — 72
    4.1.2.1 Maintenance — 73
    4.1.2.2 Cost Per Unit Water Treatment — 73
4.2 Silver-Impregnated Activated Carbon — 74
  4.2.1 Cost Considerations — 79
4.3 Electrochlorination Systems — 79
  4.3.1 Byproduct Formation — 80
  4.3.2 Cost Considerations — 82
4.4 Chlorinators — 83
  4.4.1 Liquid Chlorine as a Disinfectant — 83
  4.4.2 Chlorine Tablets — 84
  4.4.3 Disadvantages of Chlorination — 85

Keywords: UV, Humanitarian assistance, Disaster relief, Validation factor, LED, Granular activated carbon

4.1 UV LIGHT SYSTEMS

UV disinfection of potable water for humanitarian assistance/disaster relief (HA/DR) use is a viable technique to reduce or eliminate microbes present in the source water. Chemical oxidants such as chlorine often do not effectively kill spore-forming pathogens like *Giardia* and *Cryptosporidia*. Waters that do not go through filtration can contain large numbers of *Giardia* and *Cryptosporidia*; UV disinfection may be the best method to kill/ inactivate them. Many commercial UV units that purify water can be used for HA/DR applications. In Europe, as of 2001, there are more than 6000 UV systems that treat municipal drinking water (Bolton and Cotton, 2008). In the United States, there are probably more than 100 utilities using UV disinfection (Wright et al., 2009). This implies that UV systems are reliable and have already been accepted as part of municipal water systems.

☆"To view the full reference list for the book, click here"

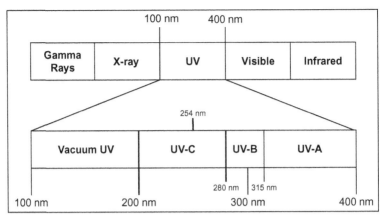

Figure 4.1 UV light in the electromagnetic spectrum (USEPA, 2006).

In the electromagnetic spectrum (Figure 4.1), UV light occupies the wavelength range from 100 to 300 nm (USEPA, 2006). Within this wavelength range, UV light is classified into four sub-spectrums: vacuum UV (from 100 to 200 nm), UV-C (from 200 to 280 nm), UV-B (from 280 to 315 nm), and UV-A (from 315 to 400 nm). UV light between 200 and 300 nm, which covers the UV-B and UV-C bands, is the most efficient at killing germs.

When electric current is applied to a gas mixture, the voltage gradient produces a discharge of photons. The wavelength of the emitted light depends on the gas mixture as well as the power level of the lamp used. Mercury is the most common element used in UV lamps. The electric current created by switching on the lamp excites the Hg atoms to a higher energy state. UV light is emitted when these excited Hg atoms return to their ground state. Xenon gas can also emit light in the germicidal range. Mercury vapor lamps are typically classified as low-pressure (LP) (or low-pressure high-output [LPHO]) and medium-pressure (MP). LP lamps produce monochromatic light at the 254-nm wavelength, whereas MP lamps produce polychromatic light between 200 and 300 nm is produced in MP lamps (Table 4.1). As shown in this table, MP lamps operate at very high temperatures, use a more electrical energy, and also produce significantly more light compared to LP lamps. However, an LP lamp has a longer operating life. Bolton and Cotton (2008) report longer operating lives for LP lamps as well as smaller power draws. According to them, the life of a LP lamp is between 8000 and 12,000 h (7000-10,000 h for LPHO lamps), and the life of a MP lamp varies between 3000 and 6000 h. The power draws for the LP, LPHO, and MP lamps, according to Bolton and Cotton, are 0.2-0.4, 0.6-1.2, and 125-250 W/cm, respectively.

Table 4.1 Typical Characteristics of Mercury Vapor UV Lamps

Parameter	Low-Pressure	Low-Pressure High-Output	Medium-Pressure
Germicidal UV light	Monochromatic at 254 nm	Monochromatic at 254 nm	Polychromatic, including germicidal range (200–300 nm)
Mercury vapor pressure (Pa)	Approximately 0.93 (1.35×10^{-4} psi)	0.18–1.6 (2.6×10^{-5} to 2.3×10^{-4} psi)	40,000–4,000,000 (5.80–580 psi)
Operating temperature (°C)	Approximately 40	60–100	600–900
Electrical input (watts per centimeter [W/cm])	0.5	1.5–10	50–250
Germicidal UV output (W/cm)	0.2	0.5–3.5	5–30
Electrical to germicidal UV conversion efficiency (%)	35–38	30–35	10–20
Arc length (cm)	10–150	10–150	5–120
Relative number of lamps needed for a given dose	High	Intermediate	Low
Lifetime (hours [h])	8000–10,000	8000–12,000	4000–8000

Note: Information in this table was compiled from UV manufacturer data.

Sharpless and Linden (2001) show that the output from LP lamps is mostly concentrated around 254 nm, and that the output at this wavelength is the maximum output. These lamps do not emit UV light at other wavelengths. However, for MP lamps, the lamp output ranges from 200 to 400 nm, with most wavelengths between 240 and 370 nm. The maximum output of a MP lamp is at 270 nm (Figure 4.2).

The UV outputs from these two different lamp types also vary over time (Figure 4.3). As the lamps age, the outputs from the lamps decrease. Solarization is a process in which electricity reacts with the mercury in a UV lamp, causing small amounts of mercury and tungsten to be deposited on the inside

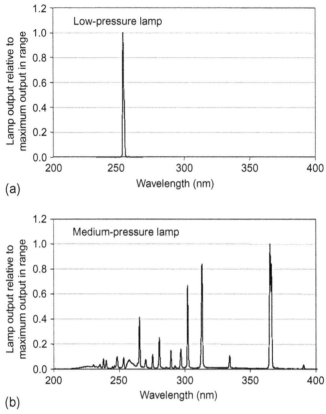

(a)

(b)

Figure 4.2 Output from (a) low-pressure and (b) medium-pressure mercury vapor lamps as functions of wavelengths. *From USEPA (2006). Source: Sharpless and Linden (2001).*

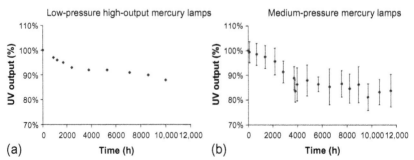

(a)

(b)

Figure 4.3 Reduction in UV output of (a) LPHO and (b) MP lamps over time. *From USEPA (2006). Source: (a) Adapted from WEDECO, (b) adapted from Linden et al. (2004).* (For the color version of this figure, the reader is referred to the online version of this chapter.)

surface of the lamp's glass envelope. This process changes the transmittance of quartz because of photo-thermal damage (USEPA, 2006). Manufacturers often provide guidelines for replacing the lamps; most of these guidelines are based on the number of hours of use. As shown in Figure 4.3, the percent reduction in output from MP lamps is much higher than that of LP lamps for same amount of time. However, remember that the output from MP lamps is much higher than that of LP lamps. The output from MP lamps seems to decrease with time in all wavelength ranges.

UV lamps typically have sleeves and ballasts. Ballasts are devices on lamps to limit the amount of current in the electrical circuit (similar to those used in fluorescent lamps to limit the amount of current through the tube). Sleeves protect the lamp from contact with water and allow the lamp to work at proper temperatures. High quality quartz is the material of choice for the sleeves. Quartz allows the transmission of UV light in the 200- to 300-nm wavelength range. The distance between the exterior of the lamp and the inside of the lamp is typically 1 cm.

When a lamp is placed in water, the emitted UV light is absorbed by the water and scattered. UV absorbance in water is a measure of UV light at 254 nm that is absorbed by water over a length of 1.0 cm. Thus, UV absorbance at 254 nm (A_{254}) is a measure of the efficiency of the UV light absorbed by water. As the wavelength of UV light increases, absorbance is decreased.

In addition to UV absorbance, UV transmittance (UVT) is another measure of UV light absorbed by water. UVT is defined as the percentage of UV light at 254 nm passing through a water sample over a distance of 1 cm.

$$\%\text{UVT} = 100 \times 10^{-A_{254}}$$

Please note that UVT and A_{254} are design parameters for UV reactors, and, in each case, the path length of light that expresses A_{254} or UVT is 1 cm. A spectrophotometer is used to measure UV absorbance as well as transmittance.

If the substance (e.g., water) has light-absorbing capacity, light gets attenuated as it travels through the medium. According to the Beer-Lambert law:

$$\frac{E^t}{E^0} = 10^{-A} = T$$

$$A = \log\left(\frac{E^t}{E^0}\right) = -\log(T) = al$$

where E^t and E^0 are transmitted and incident light radiances, A is the absorbance, T is the transmittance, a is absorption coefficient, and l is path length.

Scattering UV light can reduce its germicidal effectiveness. If the water is turbid and contains particles, the light may be backscattered toward the source (USEPA, 2006). UVT will be low in turbid waters.

UV dose is a measure of the UV intensity exposure time. The intensity is reported as mW/cm^2, and time is in seconds. Thus, a UV dose is measured in $mW\text{-}s/cm^2$, which is same as mJ/cm^2. While UV inactivation of bacteria, *Cryptosporidia*, and *Giardia* is quite effective, it is ineffective for viruses. UV inactivation efficiency is in the following order: bacteria ≈ protozoa > most viruses > bacteria spores > adenovirus > algae (Bolton and Cotton, 2008).

UV light generally inactivates microbes by damaging material in their DNA and RNA. Nucleotides such as cytosine, guanine, thymine, and adenine absorb UV light in the wavelength range 220-300 nm. After UV exposure, dimers (covalent bonds) form between two complementary nucleotides on the same DNA strand, causing damage (Figures 4.4 and 4.5).

Most damage to the DNA of infectious organisms occurs after exposure to UV light in the range of the electromagnetic spectrum. Jagger (1967; cited in USEPA, 2006) has shown that peak absorbance of UV light by DNA material occurs around 254 nm (Figure 4.6).

Bolton and Cotton (2008) state that the relative inactivation response of microbes to UV light is a measure of dimerization of the adjacent thiamine parts of the DNA strand. They define this relative response as the "inactivation action spectrum." Figure 4.7, which is redrawn using the data of Bolton and Cotton, shows that the maximum inactivation of *E. coli*, MS-2 phage, and *Cryptosporidia* occurs in the wavelength range 250–280 where the relative response is the highest in the action spectra.

USEPA (2006) also summarized similar findings that demonstrated a correlation between UV action and DNA absorbance relative to 254 nm across wavelengths from 215 to 300 nm. As can be seen, most absorbance occurs

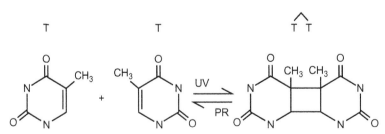

Figure 4.4 Dimerization of thiamine bases of DNA by UV light. *(ref: http://www.phys.ksu. edu/gene/e3f1.html)*

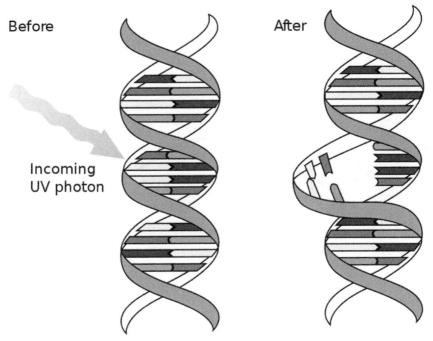

Figure 4.5 Formation of thiamine dimers on double-stranded DNA; this causes the disruption of the chain. *(ref: http://biobook.nerinxhs.org/bb/genetics/inheritance.htm)*

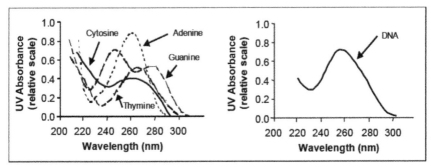

Figure 4.6 UV absorbance of nucleotides (left) and nucleic acid (right) at pH 7. *From USEPA (2006). Source: Adapted from Jagger (1967).*

between 250 and 280 nm (Figure 4.8). In MP lamps, these are the wavelengths that produce most of the UV light (see Figure 4.2).

For HA/DR applications, the sizes of UV installations can vary. For short-term use serving large population centers, the systems can be large but portable. For long-term applications, the systems should be small or

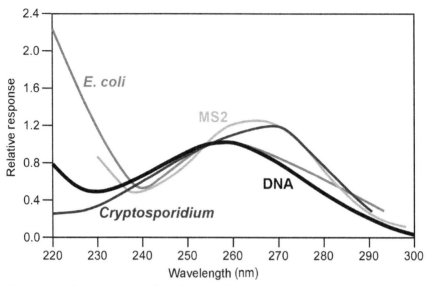

Figure 4.7 Relative response of *E. coli*, MS-2 phage, *Cryptosporidia*, and DNA to UV light between 220 and 300 nm *(redrawn using data from Bolton and Cotton, 2008)*.

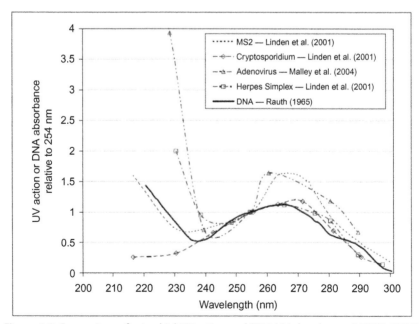

Figure 4.8 Comparison of microbial UV action and DNA UV absorbance (USEPA, 2006). Source: *Adapted from Rauth (1965), Linden et al. (2001), and Malley et al. (2004)*.

medium size. Small units can be purchased directly from the vendors and can be used to purify water for one or a few households.

SteriPEN is one of the companies that makes portable water treatment devices, including UV light systems. These units are ideal for emergency preparedness and military, travel, and outdoor recreation uses. The SteriPEN system is used for batch treatment: the UV light is inserted directly into a 0.5- or 1.0-L bottle. The unit is inserted into the neck of the bottle, and the rubberized tapered end of the unit seals the bottle neck. Then the bottle is turned upside down and shaken gently (Figure 4.9). The water-sensing pin on the unit turns on the UV lamp. The timer of the unit is calibrated to the size of the bottle, and the user may be required to press it once or twice. The light will automatically go out after a fixed period (typically 60 s for 0.5 L and 90 s for 1.0 L). The SteriPEN can also be used to disinfect water in a cup or glass. The Ultra model is sold for $99.95, and other models are cheaper. The power used to charge the unit can come

Figure 4.9 Sterilization of water in a bottle with SteriPEN (www.steripen.com). (For the color version of this figure, the reader is referred to the online version of this chapter.)

from a power outlet, a computer USB port, or from a solar panel. The unit uses two 3.0 V lithium ion rechargeable batteries. The company's web site (http://www.steripen.com/ultra/) states that use of this product results in a 3-log reduction in bacteria, viruses, and protozoa in the treated water.

Small flowthrough UV systems can be employed for household use. The U.S. National Sanitation Foundation (NSF) certifies UV systems for water treatment, and these systems are slightly more expensive than the standard UV systems. The NSF guideline requires that point-of-use (POU) and point-of-entry (POE) devices for nonpublic water systems use two optional classifications:

(a)   Class A systems: These systems use higher amounts of power (40 mJ/ cm^2) and are capable of disinfecting/removing microbes such as bacteria and viruses to safe levels from contaminated waters, and

(b)   Class B systems: These systems use lower amounts of power (16 mJ/cm^2) and are designed to provide additional bactericidal treatment of public drinking water or other drinking water (in this case, the input water to units has been already processed by a local health or regulatory agency).

The cost of NSF-certified POU/POE systems range from about $300 to $3700 (http://www.freshwatersystems.com/c-829-nsf-certified-uv-systems.aspx). For example, the Sterilight SV5Q-PA Silver series models are designed for a maximum flow rate of 3.6 gallons/min and use a 110-V power source. These units cost around $329. The lamp life is estimated to be about 9000 h, and the power consumption is 30 W. The disinfection chamber is made of stainless steel, and the unit is 22 in. long and 2.4 in. in diameter with ¾-in. MPT inlets and outlets. The system is rated as Class B (primarily for bactericidal treatment). Trojan Technology's UVMAX B4-V is a Class B unit that can handle a flow rate of 4.3 gallons/min; it costs about $350. The chamber is 14 in. long and 4 in. in diameter. Power consumption is 36 W from a 120-V source. Trojan's UVMAX Pro50 light commercial system is a Class A system for light commercial or multifamily use and can handle a flow rate of 50 gallons/min. It costs around $3675. The inlets and outlets use 2-in. MPT, and the chamber is 41 in. long and 4 in. in diameter. The power supply is from a 120-V AC source, and power consumption is 230 W.

Prices for standard UV systems (that are not certified by the NSF) are generally less than $50 for flow rates less than 1 gallon/min. This amount does not include the price of ballasts, which cost between $30 and $50. The lamp power is about 4 W, and the power is from a 120-V AC source. These units are typically small (10 in. long and 2 in. in diameter). Companies such as Pentek, Microfilter, Ultra-Sun Tech, Everpure, PURA, Trojan,

Sterilight, and Atlantic UV Corporation make such systems. Most units cost between $100 and $200, with flow rates ranging from less than 1 to 3 gallons/min. These companies also offer larger units that can be used to treat water for small communities or towns.

For villages or small towns, medium-size units can be designed using local materials. Locals only need to purchase the UV lamp, the ballast, and other essential parts from the vendors. Figure 4.10 shows the construction of LP and MP UV lamps. However, many commercial vendors also sell such lamps.

Microbes are sensitive to UV light at a given wavelength or narrow band of wavelengths, which determines the response curve of these organisms. For this purpose, the collimated beam apparatus is used to define the

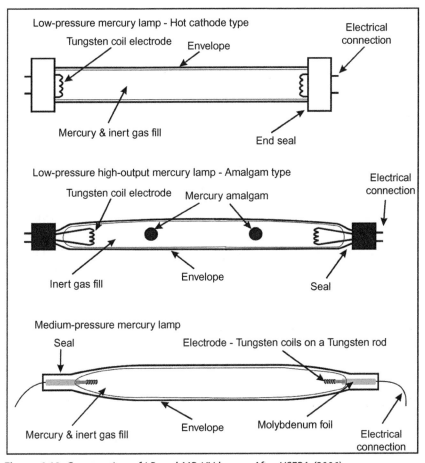

Figure 4.10 Construction of LP and MP UV lamps. *After USEPA (2006).*

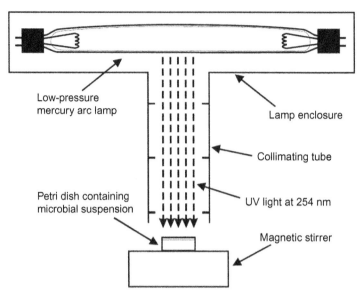

Figure 4.11 Collimated beam apparatus for developing a dose-response curve for a challenge organism (USEPA, 2006).

relationship between the UV dose and microbial inactivation. The collimated beam reduces the loss of UV light due to internal reflection of the glass surface holding the microbe samples. A suspension of microbes is left in an open petri dish, and the sample is exposed to a UV light source through a long tube so that the light beam is almost parallel and falls perpendicular to the petri dish's surface. The samples in the petri dish should remain suspended during the exposure time. Commercial vendors sell collimated beam apparatuses. It is also possible to build one without much difficulty. Thus, a homemade unit can be used as a substitute for a commercial collimated beam apparatus (see fig. 3.6, Bolton and Cotton, 2008). Figure 4.11 is a diagram of a benchtop collimated beam apparatus (USEPA, 2006).

One key step in validating the effectiveness of a UV reactor is checking biodosimetry. A challenge microorganism[1] (such as bacteriophage MS2 and Bacillus subtilis spores) is selected, and the log inactivation of the organism is measured as a function of UVT, UV intensity, and flow rate of a full-scale reactor. The dose-response curve of the challenge microorganism is developed from bench-scale testing that uses a collimated beam apparatus to plot

---

[1]According to EPA, the non-pathogenic bacteriophage MS2 and Bacillus subtilis spores are currently recognized as the standard challenge organisms for reactor validation because they are more UV resistant than most waterborne pathogens.

log inactivation versus UV dose. USEPA (2006) recommends three steps for validating UV effectiveness. Log-inactivation values from full-scale testing are input into the laboratory-derived dose-response data to estimate the reduction equivalent dose (RED), and adjustments to the RED values are made due to uncertainty and bias, which produces the validated UV dose of the reactor for the specific operating conditions. Figure 4.12 from USEPA (2006) provides the details of the validation protocol.

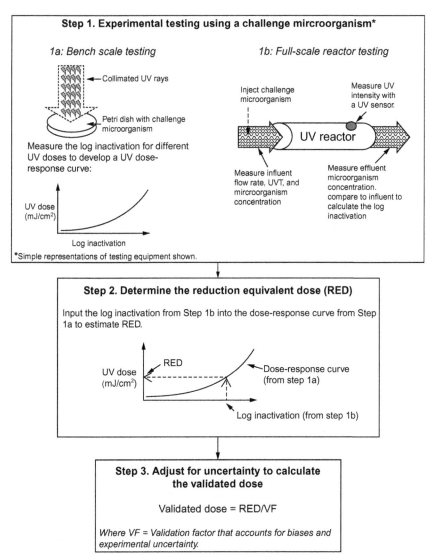

Figure 4.12 Overview of USEPA validation protocol (USEPA, 2006).

As shown in Figure 4.12, the first step is to figure out the log-inactivation values versus dose-response curve from the bench-scale testing. Then, the log-inactivation data developed from full-scale testing (step 1b) can be inserted into the dose-response curve developed in step 1a to determine the RED. The validation dose is equivalent to the RED divided by the validation factor (VF), which is >1 as it accounts for biases and experimental uncertainties. USEPA (2006) shows detailed procedures used to calculate VF and validation dose through two examples (see Appendix B of USEPA, 2006).

The hydraulics of the reactor affect the delivery of the UV dose. For plug-flow conditions (an ideal reactor), the UV dose is delivered in a narrow band but at a higher occurrence probability than a reactor with poor hydraulic conditions in which a lot of mixing occurs (Figure 4.13). Reactors with narrow dose-distribution characteristics are more efficient than reactors with wider dose-distribution characteristics. Figure 4.14 shows the dose requirements

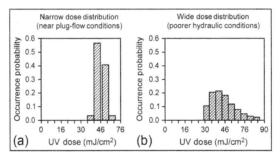

Figure 4.13 Hypothetical dose distributions for reactors with (a) good (e.g., plug flow) and (b) poor hydraulic conditions. (USEPA, 2006).

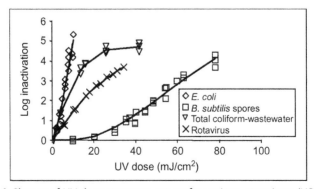

Figure 4.14 Shapes of UV dose-response curves for various organisms (USEPA, 2006). Source: *Adapted from Chang et al. (1985).*

for various organisms, *B. subtilis* spores being most resistant and requiring higher UV doses. From this figure, we can also see that Rotavirus needs a higher UV dose for the same log inactivation than *E. coli* and *B. subtilis* needs even higher dose than Rotavirus for the same log inactivation.

Recent literature suggests that microbes can also repair UV-damaged genetic materials; this is often called *photorepair* or *dark repair* (USEPA, 2006). Most bacteria can repair damaged DNA to some extent in the dark. Some of the repair mechanisms include (a) replacing UV-damaged nucleotides by removing thiamine dimers and (b) combining undamaged regions in replicated DNA molecules. Photoreactivation is another repair mechanism in which the light-activated enzyme photolase splits the thiamine dimers, thus restoring the original structure of the DNA (Dulbecco, 1949, 1950; cited in Bolton and Cotton; Kelner, 1949). The absorption of UVA light triggers this reactivation process. Photoreactivation issues are often ignored in the validation of UV reactors. However, Bolton and Cotton (2008) show details of the UV dose requirements at 254 nm for 4-log inactivation of various bacteria. *E. coli* ATCC 11229 needs only 10 mJ/cm^2 without photoreactivation compared to 28 with photoreactivation. For *E. coli* O157: H7, the required doses of UV light with and without photoreactivation are 6 and 25 mJ/cm^2, respectively. For *Cryptosporidium* and *Giardia* cysts, the doses without photoreactivation are <10 mJ/cm^2. However, for viruses, the needed doses are much higher without photoreactivation. For example, the adenoviruses require >100 mJ/cm^2 of UV. Even bacteriophages such as PRD1 or MS-2 require 30 and 62 mJ/cm^2, respectively. Typically, these doses will be higher with photoreactivation.

A substantial amount of the damage caused by LP lamps can be repaired, while most of the damage caused by MP lamps cannot be repaired. Up to 80% of pyrimidine dimers induced by LP lamps in *E. coli* can be repaired by photoreactivation (Oguma et al., 2001), while almost no photorepair of pyrimidine dimers is observed after use of MP mercury lamps, indicating that the broader range of wavelengths reduces subsequent photoreactivation (Oguma et al., 2002; Zimmer and Slawson, 2002). Therefore, it is suggested that the wavelengths produced by MP mercury lamps are critical in inactivating enzymes that catalyze photoreactivation (Kalisvaart, 2004; Zimmer and Slawson, 2002), but the precise effective UV spectrum is not known (MP mercury emission spectrum gives hints, though). This issue has not been addressed by UV-light emitting diode (LED) technology.

LED-based UV lamps are slowly becoming available for water and wastewater disinfection. Semiconductor-based Deep UV(DUV)-LED technology is becoming increasingly popular in the field of water purification because of its design freedom, low power needs, wavelength tunability, extended lifetime, and environmental friendliness (Bilenko et al., 2010; Bowsker et al., 2011). Compact DUV lamps based on Ill–Nitride semiconductors (GaN, A1N, InN, and their alloys) are now available, although they are low power (100 mW or less; see Sensor Electronic Technology of Columbia, SC). These lamps offer power distribution over a narrow wavelength (<12 nm) between 227 and 340 nm (see Bilenko et al., 2010).

Deep UV semiconductor light emitting diodes (LEDs) are mercury free and can be powered by solar panels or batteries because of the low voltage requirement (6 V). These DUV LED lamps emit between 245 and 365 nm. The best wavelength for water disinfection using DUV lamps is between 270 and 280 nm in a 3-in. diameter by 6-in. long chamber with four DUV LED lamps at each end of the cylindrical chamber. A prototype reactor created by Bilenko et al. (2010) was effective for inactivating both E. coli and MS2 1.6 logs. Kenissl et al. (2010) also tested the efficacy of UV-LED units operating at 269 and 282 nm wavelengths for flowthrough dynamics. The challenge organism was B. subtilis, and the solutions were DI water, tap water (TW), surface water (SW), and secondary effluent (SE). At 269 nm, the researchers observed that most log reduction occurred at irradiations of about 150 J/m^2 for all except DI water. With DI water, log reduction continued linearly to 400 J/m^2 and then at a slower rate to 600 J/m^2. However, at 282 nm, log reductions increased (especially in DI water) all the way to 600 J/m^2. Figure 4.15 shows the UV dose versus log inactivation for

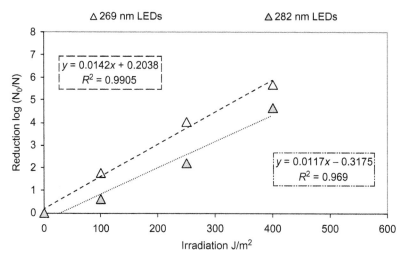

Figure 4.15 UV dose versus the inactivation of B. subtilis spores at two wavelengths (Bilenko et al., 2010).

B. subtilis spores at 269 and 282 nm. It seems the log inactivation was about one log better at 269 nm.

Kenissl et al. (2010) also tested the inactivation of *B. subtilis* in a flow-through reactor using DI water. Between 350 and 850 J/m^2, log reductions varied from <2 to 5. The researchers stated that one of the drawbacks of the system is the long exposure time required to treat water. Instead of a few seconds, an exposure time of 3-4 min was necessary. For flowthrough devices, obtaining 3-4 min of exposure time is difficult, especially in household operations.

4.1.1 Advantages and Disadvantages

Disadvantages of UV light systems are that they depend on a constant energy source, little residual disinfecting power is left after treatment, and it is difficult to accurately measure and apply the correct dose of UV. If UV-treated water is to be stored, then chemicals should be added as a disinfectant in order to prevent regrowth of microbes. Another disadvantage is that mercury lamps must be disposed of properly because of the toxicity of Hg; therefore, further investment in UV-LED lamps is recommended, although these lamps are not yet as affordable and effective as mercury lamps. There are also potential risks in UV water treatment, such as exposure to mercury if using a mercury lamp and exposure to UV light from the device, all of which are considered to be minimal. However, for HA/DR applications using small units, there is no standardized mechanism that measures or calibrates how well the equipment is working before and after being installed or throughout treatment.

The advantages of a UV light system are that it is a physical method rather than a chemical one (and thus does not leave any byproducts), it is extremely effective against protozoa, it is inexpensive, and it can treat drinking water relatively quickly. A UV light system also is easy to install and requires little maintenance, which helps keep costs low. UV treatment does not produce any kind of chemical taste or smell and is not sensitive to pH or temperature. It does not take any minerals out of the water, improves taste because it destroys some organics, requires little contact time, has no smell, has no volatile organic compound emissions, and was recently acknowledged to control *Cryptosporidium*.

UV light can be used as a primary treatment system, and this method has been used in North America and Europe for some time (Palaez, 2011). UV systems can be used alone or with other types of treatments. For example, it is common to use both UV light and ozone/hydrogen-peroxide for groundwater treatment. UV water treatment systems are sometimes produced by companies that also manufacture ozone treatment systems. Chemicals such as chlorine destroy cell structures, interfere with metabolism, and hinder biosynthesis and growth. In contrast, UV light damages a microorganism's nucleic acid, which hinders its replication, and a microorganism can't infect a host if it can't reproduce.

4.1.2 Design Considerations

UV light with an intensity of 90-120 mW/cm^2 reduces *Giardia lamblia* by 1-log, and UV light with an intensity of 90-140 mW/cm^2 reduces viruses by 4-log. Because there is no standardized measure to check the effectiveness of UV light systems, water is frequently tested to ensure effluent has low counts of total coliform and HPC (heterotrophic plate count) bacteria. The UV spectrum for disinfection that is most successful and most practical is the UV-B range of 280-315 nm. This range has the highest germicidal action, although UV-C of 200-280 nm can also be used with longer exposure times.

The dose of the UV light must be high enough that the organism's DNA and RNA will be disrupted. This depends on the lamp intensity (the rate that the photons are delivered to the target), selected wavelength, radiation concentration, water quality, flow rate, exposure time, type of microorganism, and distance to the lamp. Contact time is difficult to measure and is influenced by the flow rate and the reactor hydraulics. However, UV light system manufacturers conduct biodosimetry tests under controlled laboratory conditions and supply the resulting contact time recommendations to purchasers.

Depending on the type of system, UV lamps can run on batteries, solar panels, and electricity. The power supply is important and must be reliable; in situations where power is not reliable, a generator is recommended. As stated earlier, most UV drinking water treatment uses LPHO or MP lamps, which have a lifetime of about 8000-12,000 and 4000-8000 h, respectively. LPHO lamps are more power efficient and run at a lower cost than MP lamps.

Each system should have a flow rate control that allows for a maximum flow that relates to the size of the reactor. A minimum velocity of water is needed to prevent fouling, as mentioned above. This velocity depends on the size of the reactor. Flow rates should be between the minimum and maximum, which should be clearly identified. A minimum flow rate also prevents overheating of the lamps.

To ensure that the challenge protozoa are not a concern, it is best to use a functional system that was previously built and tested instead of building a system locally. This allows users to have confidence in the system. Users can also install two units so that one unit can treat water while the other is being cleaned or maintained; this will also be helpful in low-flow situations. Most preassembled systems come with automatic cleaners and alarm systems, which are easy to replace.

In order to prevent biofouling, the UV light reactor should remain on, and the system should never be left full of stagnant water for an extended period, or a biolayer may form. If users follow the recommended cleaning protocols and wipe the lamp surface 1-12× an hour, fouling can be significantly decreased. Mineral or inorganic fouling is difficult to predict and stop; this process is rather complicated, and it is recommended that users properly clean and maintain the reactor for the best results and to increase the life expectancy of the system. Another way to reduce biofouling is to maintain a constant flow rate that is appropriate for the system. If the flow rate is too low, a biolayer can begin growing, and, if too fast, the light does not have enough contact time with the water to disinfect it.

Knowing the pH, alkalinity, hardness, iron (Fe) content, and magnesium (Mg) content of the water before it reaches the UV reactor will help determine what kind of fouling will occur and how best to resolve the effects. These factors do not affect the disinfection process itself, only the obstruction of the UV light when a crust forms. For example, one wiper sweep every 15 min for a total calcium and hardness of less than 140 mg/L and with an iron of less than 0.1 mg/L.

4.1.2.1 Maintenance

UV units are simple to maintain. The lamps should be replaced yearly or when efficiency decreases to 70%. Users should inspect the lamps often, clean them every 6 months, and replace the O-rings, valves, switches, and ballasts as needed. An operator should monitor these systems and carefully watch the water's turbidity and color because these characteristics relate to the system's efficiency. Users should also carefully monitor the water's calcium content because it also has a negative effect on the system's performance.

4.1.2.2 Cost Per Unit Water Treatment

Costs for operating and maintaining UV light water treatment systems vary from $0.02 to $2.35 per m^3 (Palaez, 2011). These systems are affordable for low-income households and communities, and these systems can be used in environments with minimal electrical and water infrastructure.

Initial investments vary, depending on the size of the units and the addition of monitoring units. The estimated cost of a household system can be as low as $50; for use in a small community, a high volume unit that incorporates the most up-to-date technology can cost as much as $86,419. The average cost for a community-size UV water treatment system is between $300 and $900 (Palaez, 2011).

4.2 SILVER-IMPREGNATED ACTIVATED CARBON

Silver in various forms has been used as an antibacterial agent for centuries. Silver is used as a biocide in water treatment as well as in other applications such as dyes and paints, varnishes, textiles, cosmetics, cleaning agents, baby bottles, washers, and more. For example, the "Barbicide" used in barber shops contains silver nitrate along with alkyl dimethyl benzyl ammonium chloride. Use of silver nitrate as a germicidal agent dates back to the late 1800s (Kim et al., 2009). Silver in various forms is used in cartridge filtration systems that employ activated carbon to absorb contaminants. Silver is primarily used to inhibit bacterial growth on the carbon's surface.

Cartridge filtration systems are the most common point-of-use water treatment devices that use various types of filters in designated housings to produce potable water. Although home-use granular activated carbon (GAC) filters became very popular in the 1970s, these systems did not address bacterial growth (Taylor et al., 1979). Geldreich et al. (1985) showed the colonization of bacteria in POU devices based on daily use or nonuse. None of the filters used in their 3-year study contained silver impregnated activated carbon. They found daily variability in the release of bacteria from the devices and colonization of the microbes during nonuse periods. Brewer and Carmichael (1979) showed that microbial populations adhere to and accumulate on a GAC surface during water filtration and bacterial endotoxins can be transmitted to the filtered water.

Research on the use of silver-impregnated carbon filters for water treatment dates back to the 1970s in the United States. Regunathan and Beauman (1987) reported that adding silver to GAC filters reduced bacterial counts compared to filters without silver. However, in their study, the control counts were low for coliforms. They also tested filters containing copper but reported that the antimicrobial effects of copper were much less than those of silver. The GAC in these filters has a large surface area and can serve as a broad-spectrum adsorption bed. Such systems are often used in HA/DR scenarios as well as in developing countries. In HA/DR applications, the source waters are typically surface waters. For household use in developing countries, the source water is typically city water fed to the homeowner's tanks or shallow ground water obtained from a well located on the homeowner's property. There is concern about whether the silver is effective over a prolonged period as well as whether it can leak out of the filters and into the treated water.

Bell (1991) shows that contact time was crucial in reducing *E. coli* in water treated with filters containing silver nitrate or silver ions. The killing action of typical filters is not effective because the contact is limited to minutes for household filters. Bell also points out that several studies have raised questions about the efficacy of silver-impregnated activated carbon filters and that most published data have been negative about the use of silver to control bacteria. At 50 μg/L silver concentration and at 1 h contact time, *Salmonella* was reduced by 5 logs, *Pseudomonas aeruginosa* was reduced by 50%, and there was no reduction of polio viruses. Reasoner et al. (1987) described a USEPA study of three filters with silver and four without silver that were used for 11 months. The heterotrophic plate count (HPC) for the influent during this period was about 6000 per mL. However, the effluent HPC ranged from 800-62,000 for filters with silver to 32,000-107,000 for filters without silver. All these units showed an increase in the bacterial population.

Currently, many vendors provide silver-impregnated activated carbon filters for home use. These filters typically cost less for home use, and volume discounts are available. Pentek manufactures silver-impregnated GAC filters with nominal pore sizes of about 0.5 μm. These filters have a diameter of about 73 mm and a height of 248 mm, and the design flow rate is 1 gallon/min (Figure 4.16). The life of the filter is about 1000 gallons. This filter fits into a 10-in. filter housing.

Often, high-quality coconut GAC filters impregnated with silver are used in large filter casings. However, the end user must be aware that silver leaks out of the carbon during the water filtration process. Contact times and the amount of carbon are the two factors that determine the antimicrobial action of these filters. Figure 4.17a shows a filtration unit that uses nano-silver for bacterial reduction. The filter is made for household use by the Tata Company in India (http://www.tataswach.com/). This filter has a bulb in the center through which the source water passes under gravity (Figure 4.17b); the lower unit is a container to hold the filtered water. Two such units were tested in the authors' laboratory to filter settled stream water.

The silver on the top and bottom of the filter bulb after a few weeks of use is shown in Figures 4.18 and 4.19. Only a few silver nanoparticles are

Figure 4.16 Silver-impregnated activated carbon filters from Pentek® (www. pentekfiltration.com). (For the color version of this figure, the reader is referred to the online version of this chapter.)

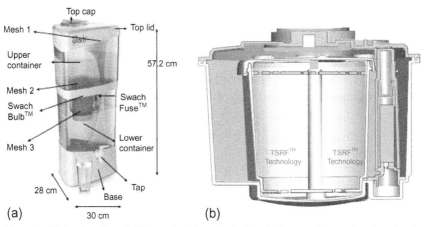

Figure 4.17 (a) Tata Swach filter and (b) its filter bulb, which contains sand and husk ash with silver nanoparticles. (For the color version of this figure, the reader is referred to the online version of this chapter.) *(www.tataswach.com)*.

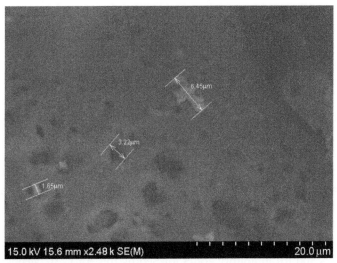

Figure 4.18 Silver nanoparticles on the husk surface on the top portion of the filter bulb (white spots).

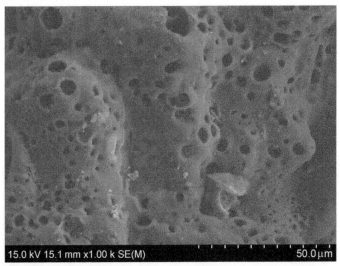

Figure 4.19 Silver nanoparticles on the husk surface on the bottom portion of the filter bulb (white spots). More nanosilver is present on the bottom as the contaminants present in the water being filtered have not reached the bottom yet.

present on the husk in Figure 4.18 compared the number in Figure 4.19. As the water is filtered through the bulb, the nanoparticles of silver present in the top section of the filter are washed away. Silver release from silver-containing antimicrobial textiles has been documented by Lorenz et al. (2012).

They investigated eight commercially available silver-containing textiles with total silver contents ranging from 1.5 to 2925 mg/kg. Four of the eight textiles released silver, and the majority of the released silver (34-80%) was in the form of particles >450 nm.

Bandopadhyaya et al. (2008) provide some details of how silver-impregnated GAC filters are prepared for water purification. They prepared $AgNO_3$ solutions (0.001-1.5 M) and added 1 g of GAC to 20 mL of $AgNO_3$. The solution was kept in the dark to reduce photodecomposition of the $AgNO_3$. After 24 h of carbon impregnation, the samples were washed with water to remove any loosely adsorbed $AgNO_3$ from the GAC. The GAC was dried on a filter paper in a vacuum desiccator. By adding 0.2 M $NaBH_4$ to the GAC, the $AgNO_3$ was chemically reduced to Ag particles. Excess $NaBH_4$ was washed away, and the GAC was dried. Figure 4.20 shows (a) scanning electron microscope (SEM) image of silver on GAC as depicted as white spots and (b) inhibitory zones formed by this Ag-GAC against *E. coli*.

As mentioned earlier, the drawback of silver-impregnated GAC is the loss of silver as water passes through the filter. Manufacturers of antibacterial carbon fibers (Oya et al., 1993; Park and Jane, 2003) have created procedures to prepare silver-containing activated carbon fibers (ACF) for antibacterial use. Zhang et al. (2004) suggest that Ag/ACF systems do not work well because the silver is normally much larger (20 nm to 1.5 μm) than the pores on an ACF (about 2 nm). Because the silver is not trapped in the pores, it is

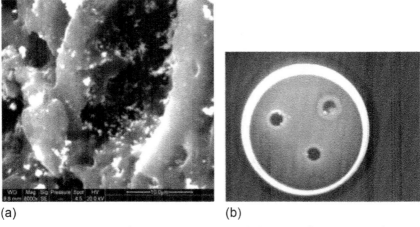

(a)   (b)

Figure 4.20 SEM image of silver (white spots) on dark GAC surfaces (a) and inhibitory zones formed by this Ag-GAC against *E. coli*. After Bandopadhyaya et al. (2008). (Used with permission from Wiley Online Library). (For the color version of this figure, the reader is referred to the online version of this chapter.)

easily washed off. Zhang and colleagues recommend using carbon aerogels (CA) as alternatives to ACF because the CAs have low bulk densities and high surface areas and are highly porous. The bulk density of a CA is typically between 0.2 and 0.4 g/cm^3, and the silver content can be adjusted as desired. Zhang and colleagues used silver nanoparticles on the CA bed, which was dried in ambient pressure. They tested the antibacterial properties of CA and found that at 1.4% Ag, 5 mg of Ag-doped CA can kill nearly all *E. coli* in a 10-mL solution at 1.92×10^5 CFU/mL in 2 h.

Srinivasan et al. (2013) synthesized Ag nanoparticles under UV light by reducing $AgNO_3$ with sodium citrate. The Ag particles had a mean diameter of 28 nm with a standard deviation of 5 nm. The activated carbon was treated with oxygen plasma to increase polar functional groups. This treatment increased the number of Ag nanoparticles on the external surface of activated carbon compared to inside the pores. FTIR spectra showed an increase in the amount of oxygen-containing functional groups such as C—O or C=O from about 22% to 31%. In their plate assay and shake flash tests, these researchers found a one-order increase in the death rate of *E. coli*.

4.2.1 Cost Considerations

A comparison of the cost of producing a gallon of water using various filters for under the sink, on the counter, or elsewhere can be found at WaterFilterComparisons.com. For nonbacteriostatic filters, the cost per gallon of water varies between 9 and 26 cents. The cost for operating countertop units varies between 12 and 16 cents per gallon. Initial investments vary depending on the brand.

The Pentek® brand 10-in. silver-impregnated GAC filter costs about $25. The housing, tubing, prefilters, and other equipment can cost another $100. The flow rate is 1 gallon/min, and the filter is designed to produce 1000 gallons of water. If the prefilters are replaced at the same time as the GAC filter, then the cost for 1000 gallons of water is around $50, which translates to about 5 cents per gallon. If the filters perform at 50-75% of the projected efficiency, the cost per gallon is about 8-10 cents.

4.3 ELECTROCHLORINATION SYSTEMS

Electrochlorination is a technique used to produce sodium hypochlorite (NaOCl) from salt water or sea water by running electric current through it. NaOCl is bleach, which is used in household applications and is also widely used in water treatment processes. The product has a pH between

6 and 7.5 and is stable. Typically, the energy source is electricity or a battery. The chemical reaction is:

$$NaCl + H_2O + ENERGY \rightarrow NaOCl + H_2$$

At anode, free chlorine is generated using the following reaction:

$$2Cl^- \rightarrow Cl_2 + 2e^-$$

At cathode, H_2 is produced in the aqueous media as follows:

$$2H_2O + 2e^- \rightarrow 2OH^- + H_2$$

In the anode reaction, Na^+ ions and Cl_2 gas are present in excess quantities. OH^- ions migrate from the cathode region to the anode region and react to form:

$$OH^- + Na^+ + Cl_2 \rightarrow NaOCl$$

The balanced chemical reaction can be written as:

$$2NaOH + Cl_2 \rightarrow NaOCl + NaCl + H_2O$$

The Cl_2 gas produced at the anode is primarily used to form NaOCl; however, H_2 produced at the cathode is lost to the atmosphere.

Either pure NaCl is added to deionized water, or seawater is used to make bleach. Several commercial companies make electrochlorination units to treat drinking water. Cascade Designs and Miox Corporation make small units that have been used for HA/DR applications (Figure 4.21).

4.3.1 Byproduct Formation

If seawater is used for electrochlorination, bromate may be created:

$$3OCl^- + Br^- = BrO_3^- + 3Cl^-$$

More recently, electrode materials have been investigated for their ability to produce different types of oxidants for disinfection. Jeong et al. (2009) examined boron-doped diamond (BDD), Ti/RuO_2, $Ti/Pt-IrO_2$, and Pt as electrodes. The efficiency of $_OH$ production, as determined by para-chlorobenzoic acid (pCBA) degradation, was, in order: $BDD \gg Ti/RuO_2 \approx Pt$. No significant production of $\dot{}OH$ was observed at Ti/IrO_2 and $Ti/Pt-IrO_2$. The $\dot{}OH$ was found to play a key role in O_3 generation at BDD, but not at the other electrodes. The production of active chlorine was, in order: $Ti/IrO_2 > Ti/RuO_2 > Ti/Pt-IrO_2 > BDD > Pt$. Li et al. (2011) compared electrochemical disinfection with ozone treatment, chlorination, and monochloramination of treating drinking water. They found

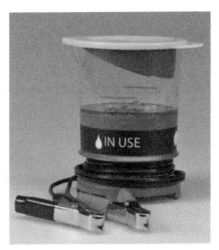

Figure 4.21 An electrochlorinator made by Cascade Design, Inc., that can generate enough NaOCL to treat 200 L of water in a few hours (http://sites.path.org/water/files/2011/09/SE200-super-money-shot1-300x334.jpg). (For the color version of this figure, the reader is referred to the online version of this chapter.)

that, under optimal conditions, CT (concentration of disinfectant × contact time) values of a 4-log *E. coli* inactivation were 33.5, 1440, 1575, and 1674 mg min/L for the electrochemical process, ozonation, chlorination, and monochloramination, respectively. Dalaedt et al. (2009) evaluated the disinfection of *E. coli* and *L. pneumophila* in tap water (which was seeded with these organisms) using an electrochemical cell (Ecodis®) with an internal volume of 1 L. Disinfected tap water was used in the electrolysis cell to generate the oxidant. The small Ecodis cell was able to disinfect 150 L of water in an hour. The free oxidants in the water were measured using a Hach® test model DR/3 and Hach DPD free chlorine reagent powder pillow. The chloride in the tap water was sufficient to produce enough free chlorine for disinfection.

Jeong et al. (2007) report that there has been tremendous interest in recent years in the electrochemical disinfection of water. Most efforts have been geared toward generating free chlorine electrochemically, and research on oxidant production without chlorine is scarce. Jeong et al. mention that disinfection occurs in two stages: (a) electrosorption of negatively charged microbes to the anode surface, followed by (b) a direct electron transfer reaction. As electrolysis continues, the inactivation rate becomes slower than in the first phase but remains steady. Hydroxyl radical plays a significant role in disinfection. Although these oxidants do not leave significant residuals, they also do not form other byproducts such as bromate when seawater is used for electrolysis.

Studies show that perchlorate and bromate are produced by electrolysis of seawater (Oh et al., 2010). Perchlorate concentrations were low compared to bromate concentrations. Bromate is a regulated compound, and the current USEPA standard for bromate in drinking water is 0.01 mg/L annually. The molecular weight of bromine is 79.9 g/mol, and that of hypochlorite ion and bromate are 51.5 and 127.9 g/mol, respectively. A typical concentration of Br^- in seawater is 67.3 mg/L (0.84 mM) at 3.5% salinity. Also, 3 mol of hypochlorite ions react with 1 mol of bromide to form a bromate ion. Thus, 0.84 mM (107.7 mg/L) bromide can theoretically produce 0.84 mM bromate if sufficient OCl^- is present. This can far exceed the MCL.

Many large manufacturers produce units for electrochlorination/oxidation. Anodic oxidation (as it is often called) is frequently used in these devices. Hydrosystemtechnik GmbH of Bruckmuehl, Germany, produces such units, which have been deployed in various locations. Kraft et al. (1999a) compared the production of hypochlorite from dilute chloride solutions using platinum- and iridium oxide-coated titanium expanded metal electrodes used as anodes. They found that the titanium electrode coated with iridium oxide produces hypochlorite at a higher rate than the titanium electrode coated with platinum. Kraft et al. (1999b) also found that the production of hypochlorite dropped off with increased temperature, but an increased rate of production was associated with higher rates of flow through the electrode assembly. In addition, Kraft et al. (2002) stated that calcium coating on the cathode reduced the efficiency of disinfection. They suggested using the cathode as a sonotrode, which can be cleaned while it is connected to an ultrasonic unit (Figure 4.22).

4.3.2 Cost Considerations

The cascade unit costs about $100, and it needs a battery or some sort of DC voltage source (i.e., a solar panel with a battery) to function. The only chemicals needed to generate NaOCl are pure salt (operators should avoid using sea salt as it contains bromide ions, which can form bromate). In India (Mumtaz et al., 2012), the cost of electrolytical defluoridation is about $0.15 per 1000 L of water. Capital costs of units that use batteries can be around $1500. A solar panel set-up for this system would cost about $2000 more.

Figure 4.22 Hydrosys AO System® RPZ 2/160 HT-KW. *Courtesy of Meinolf Schoeberl, Hydrosys, Germany.* (For the color version of this figure, the reader is referred to the online version of this chapter.)

Kalf Engineering Pte Ltd, in Singapore, makes "elysisPURE" electrochemical units that are used in water treatment operations for continuous flow systems. These devices, according to the manufacturer's web page, can produce hypochlorite with chlorine at >8000 mg/L. The electrodes last for 5 years or longer, and the unit's life is about 7 years. The power needed to produce 1 kg of chlorine is about 4.2 kWh, and the salt needed is about 3.5 kg. The unit should be cleaned with acid once a year. The elysisPURE series NT-H-PTET-B960 units produce 40 kg of sodium hypochlorite per hour at a chlorine concentration of 8000 mg/L. Kalf Engineering also has other units with chlorine production rates ranging from 1 to 2000 kg/h.

4.4 CHLORINATORS

Chlorine is an inexpensive form of water treatment. It is available in several forms and can easily treat large as well as small volumes of water. Because of chlorine's versatility, chlorine disinfection devices can be used for to treat water for both households and communities (Backer, 2008; Loo et al., 2012).

4.4.1 Liquid Chlorine as a Disinfectant

Household bleach can be used to disinfect drinking water. A specific number of drops should be added to the water, depending on the concentration of the chlorine in the bleach solution, the quantity of the water being treated, and the quality of the water being treated. Water quality factors that increase the necessary chlorine dose include increased turbidity, increased pH, and decreased temperature (Burch and Thomas, 1998). As turbidity impedes

Table 4.2 EPA (2006) Recommendations for Disinfecting Contaminated Water Using Household Bleach

Available Chlorine	Drops Per Quart-Gallon of Clear Water	Drops Per Liter of Clear Water
1%	10 per quart-40 per gallon	10 per L
4-6%	2 per quart-8 per gallon	2 per L
7-10%	1 per quart-4 per gallon	1 per L

Double the amount of drops if water is murky or turbid.

disinfection, water should be filtered using a cloth filter or a coffee filter before chlorination. Inactivation of microorganisms such as *Vibrio*, *Salmonella*, and *Shigella* generally requires a contact time of 30 min, depending on the dose used. Chlorine has been found to be fairly ineffective against protozoans because they can form spores, so a much longer contact time of 4 h is required. Even with the longer contact time, chlorine might still be ineffective against some microbes, such as *Cryptosporidium* (Table 4.2).

An advantage to using chlorine as a disinfectant is that it leaves chlorine residual in the water. This residual helps prevent recontamination of the water if it's stored correctly. Chlorination using sodium hypochlorite (NaOCl) was found to improve stored water quality after the Indonesian tsunami (Gupta et al., 2007; Loo et al., 2012). NaOCl can also be used for cleaning storage vessels, but, in some cases, it did not prevent recontamination of storage vessels (Steele et al., 2008).

When water is treated as a batch solution with chlorine drops, little maintenance of the treatment device is required. Filtering the water and adding the chlorine drops does not require skilled operation. Chlorine degrades quickly, so the treated water needs to be stored in a closed container in a cool, dark place so the degradation process is slowed. Even then, the half-life of chlorine is 2 months. Because of this quick degradation, chlorine supplies need to be replenished often to ensure sufficient disinfection. The costs associated with chlorine use are mainly derived from the cost of replenishing the chlorine. Labor takes around 20 min to produce a batch of disinfected water and thus is not highly time sensitive (Burch and Thomas, 1998). However, if turbid water must be filtered, costs and labor will increase.

4.4.2 Chlorine Tablets

Loo et al. (2012) and Berg (2010) noted that chlorine in tablet form such as sodium dichloroisocyanurate ($NaC_3N_3O_3Cl_2$, abbreviated as NaDCC) has been widely used in emergencies, and studies have shown NaDCC to offer advantages over liquid chlorine, such as greater stability, safety, and convenience because of its single use packaging and light weight (Clasen and Edmondson, 2006; Lantagne et al., 2010). Tablets also have a longer shelf life and lower transportation costs than liquid chlorine. McLennan et al. (2009) compared four types of disinfection POUs for emergency water treatment and found that none of the disinfection technologies, NaDCC included, provided a large enough chlorine residual for safe water storage.

Similar to liquid chlorine, the dosage of chlorine tablets depends on factors such as temperature, turbidity, presence of NOM (natural organic matter), and the type of bacteria/viruses (Abbaszadegan et al., 1997; Feachem et al., 1983; LeChevallier et al., 1981). NaDCC tablets have been shown (Clasen et al., 2007; Jain et al., 2010) to reduce risk of diarrhea by 48% in Zambia. However, in a recent epidemiological study in Ghana, no significant reduction in diarrheal risk was associated with using NaDCC tablets (Jain et al., 2010). These conflicting findings regarding diarrhea reduction could be attributed to different levels of water contamination as well as the procedures used to store the water (Loo et al., 2012).

Loo et al. (2012) also noted another form of available chlorination technology: water-insoluble polymeric beads that release halogen when in contact with microbes (Chen et al., 2003; Mazumdar et al., 2010). The disinfection mechanism is probably a diffusion-induced release of halogen. These biocidal beads can be integrated into filtration processes for controlled flow rates and predictable performance. Nevertheless, NaDCC tablets are the preferred chlorination technology because these tablets are easy to handle and rapidly deployed during acute emergencies.

Aquatab manufactures NaDCC tablets for emergency disaster relief. The NaDCC dose and the corresponding amount of water that it will treat are shown in Table 4.3. Tablets are dropped in water, and the water is stirred and then allowed to sit for 30 min.

4.4.3 Disadvantages of Chlorination

A disadvantage of using chlorination to disinfect drinking water is the potential formation of harmful byproducts if too much chlorine is used and there are organic compounds in the water. Chlorine byproducts can form carcinogens and trihalomethanes (THM), but a study by Lantagne (2010) found that THM levels created by NaDCC tablets and NaOCl were below acceptable levels specified by World Health Organization.

Table 4.3 Aquatab Concentrations for Emergency Response
Applications (Medentech)

Concentration	Quantity of Water to Be Treated
Aquatabs 8.5 mg	1 L
Aquatabs USA	0.8 quarts
Aquatabs 33 mg	5 L
Aquatabs 67 mg	10 L
Aquatabs 167 mg	20 L
Aquatabs 1.67 g	200 L
Aquatabs 8.68 g	1000 L
Aquatabs Granules	All volumes greater than 1000 L

Another disadvantage of chlorination is that chlorine is ineffective against *Cryptosporidium* and *Mycobacterium* (Backer, 2008; Clasen et al., 2007), both microbes that can cause illnesses.

For all treatment technologies that include chlorination, the required dose depends on the amount of water being treated. Without knowing the size of water container, it's difficult to know how much chlorine to add, resulting in the risk of overdosing or underdosing the water. A study by Clasen et al. (2007) found that a 67-mg tablet was an appropriate dose in a 20- to 25-L water container, but this size tablet was too much chlorine for a 12- to 14-L container. It's interesting to note that the values Clasen thought were too high do not correspond with the values in Table 4.3, the dose values suggested by the manufacturer. The dose also depends on the water's turbidity. The authors agreed that overdosing was less risky than underdosing, which would result in insufficient disinfection. However, overdosing could result in less acceptance of chlorine disinfection because of the strong chlorine taste and smell that result from the free chlorine residuals (Clasen et al., 2007).

CHAPTER 5

Technologies for Long-Term Applications

Contents

5.1 Slow Sand Filtration	88
5.1.1 Removal Efficiency	89
5.1.2 Construction	89
5.2 Packaged Filtration Units	103
5.2.1 Candle Filter	103
5.2.1.1 Materials, Manufacturing, and Removal Efficiency	*104*
5.2.1.2 Improve Filter Efficiency	*105*
5.2.1.3 Maintenance	*105*
5.2.1.4 Cost	*106*
5.2.2 Ceramic Disk Filter	106
5.2.2.1 Materials, Manufacturing, and Removal Efficiency	*106*
5.2.2.2 Maintenance	*108*
5.2.2.3 Cost	*108*
5.2.3 Ceramic Pot Filters	108
5.2.3.1 Materials, Manufacturing, Removal Efficiency	*109*
5.2.3.2 Maintenance	*109*
5.2.3.3 Cost	*110*
5.2.4 Evaluation of Ceramic Water Filters	110
5.2.5 Lifestraw Personal	111
5.2.5.1 Cost	*111*
5.2.6 Lifestraw Family	111
5.2.6.1 Cost	*114*
5.2.7 FilterPen	114
5.2.7.1 Cost	*115*
5.2.8 Chulli (Ovens) Treatment	115
5.3 Pressurized Filter Units	116
5.3.1 Multistage Backpack Filter	117
5.3.1.1 Operating Removal Efficiency	*118*
5.3.1.2 Cost	*118*
5.3.2 Packaged and Portable RO Filter	118
5.3.2.1 Operating Removal Efficiency	*118*
5.3.2.2 Cost	*119*
5.3.3 WaterBox	119
5.3.3.1 Cost	*121*

☆"To view the full reference list for the book, click here"

    5.3.4 Lifesaver Jerrycan   121
        5.3.4.1 Cost   *122*
5.4 Small-Scale Systems   122
    5.4.1 Sunspring   124
        5.4.1.1 Cost   *125*
    5.4.2 Perfector-E   126
        5.4.2.1 Cost   *127*
    5.4.3 SkyHydrant   127
        5.4.3.1 Cost   *128*
    5.4.4 iWater Cycle   128
        5.4.4.1 Cost   *128*
    5.4.5 Evaluation of Small-Scale Systems   129
5.5 Natural Filtration   130
    5.5.1 Design of Wells   132

Keywords: SSF, Packaged filtration, Natural filtration, small scale systems, pressurized filtration

5.1 SLOW SAND FILTRATION

Slow sand filtration (SSF) is a low-cost, yet effective treatment option for removing microbes in source water. It has helped tremendously in stopping the spread of gastrointestinal diseases in developing countries. SSF was first employed for municipal use in the 1850s in Great Britain (Logsdon et al., 2002), then was implemented in other European countries and eventually the United States in 1885 (Huisman and Wood, 1974). SSF saved hundreds of lives during the 1892 cholera epidemic in Germany, particularly in Altoona, which is downstream of Hamburg (Logsdon et al., 2002). This is one of the many cases that have demonstrated the efficacy of SSF even though the science of microbiology was still developing at that time.

Untreated fresh water (from rivers, lakes, rain water, etc.) is fed to a sand bed that is typically 0.5 m in depth or deeper in some cases. The water percolates down through the sand because of gravity. A certain amount of water stays on the surface of the sand at all times, creating a pond-like environment. With time, a biolayer (often referred to as a *schmutzdecke*) grows on the top layer of sand particles that are under water. It is typically reddish brown and has a slimy film. This layer acts as a preliminary filter that removes fine colloidal particles and organic matter from the feed water (ITACA, 2005). The biolayer also helps to remove bacteria and pathogens through sorption, predation, and other mechanisms.

Table 5.1 Literature Data on the Efficacy of Slow Sand Filtration on Removal of Microbial, Chemical, and Physical contaminants in Source Water (from Gottinger et al., 2011)

Parameter	Removal Efficiency or parameter value in effluent	Reference(s)
Turbidity	<1 NTU	Visscher (1990), Galvis et al. (1998) in Cleary (2005)
True color	25-40%	Galvis et al. (1998) in Cleary (2005)
	30-100%	Visscher (1990)
Organic matter	60-75%	Visscher (1990)
UV absorbance (254 nm)	3-35%	Galvis et al. (1998) in Cleary (2005)
THM precursors	<25%	Galvis et al. (1998) in Cleary (2005)
Viruses	Almost complete removal	Visscher (1990)
Fecal coliforms	95-100% to 99-100%	Visscher (1990)
Standard plate count bacteria	96%	Bellamy et al. (1985a)
Giardia cysts	Almost 100%	Bellamy et al. (1985a)
Cryptosporidium oocysts	99.8-99.99%	Various citations in Logsdon et al. (2002)
Poliovirus	99.997% average	Poynter and Slade (1977) in Logsdon et al. (2002)
Iron and manganese	30-95%	Visscher (1990)

Note: THM = trihalomethane.

5.1.1 Removal Efficiency

Typically, water turbidity is reduced to less than 1 NTU in municipally operated SSF units. Gottinger et al. (2011) have compiled the removal efficiency statistics of SSF units for various contaminants, including microbial and chemical as well as physical impurities present in source water. In well-operated systems, a significant amount of these contaminants are removed (Table 5.1). What is striking about this data is that almost all *Giardia* cysts are removed and up to 4 log removals of *Cryptosporidium* oocysts is achieved. More than 2 logs of fecal bacteria removal is achieved. Polio virus removal of up to 5 logs is attained. Trihalomathane (THM) precursor removal is shown to be less than 25%.

5.1.2 Construction

For a municipal or public water supply, the construction of SSF is quite elaborate, and standard texts and design manuals or handbooks are available

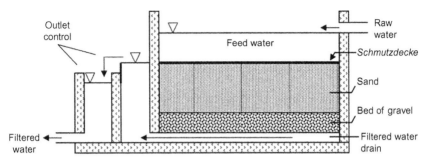

Figure 5.1 Diagram of a slow sand filtration unit used by water utilities.

(Randtke and Horsley, 2012). Figure 5.1 shows a diagram of a slow sand filter typically used to treat public water supplies.

Typically, each filter is a concrete box with appropriate pumps, pipes, underdrains, and baffles. As shown in the figure, raw water is fed to the top of the filers from one side of the wall. The sand in the filter sits above a gravel bed that acts as support as well as a drain. The water level in the filter bed is adjusted by a weir just outside the wall so that there is a small depth of water just sitting on the top of the sand bed. This allows the *schmutzdecke* to grow. The overflow from the weir is collected as filtrate. It may be further treated to comply with local, state, and federal water quality regulations for public water utilities.

The town of Falls City, Oregon, has a population of about 950. It built an SSF unit at a cost of about $1.5 million (by SCG Slayden of Slayton, OR) that included the costs of three filter beds; pumps; piping; a covered chlorine

Figure 5.2 An aerial view of the slow sand filter unit for the town of Falls City, Oregon. http://www.slayden.com/falls-city-slow-sand-filter/ (For the color version of this figure, the reader is referred to the online version of this chapter.)

contact basin and clear well; and the adjacent building for controls, monitoring, and office space (Figure 5.2). The source water comes from a creek.

While SSF has been effective in treating large amounts of raw water from streams, lakes, or ponds, its construction may take some time, and, thus, large, permanent SSF systems may not be effective in emergency situations. Immediately after disasters, portable SSF can be set up in plastic barrels or brick tanks with appropriate piping so that potable water can be produced in a week or two. Community-scale or smaller (one or a few houses) systems can be set up easily.

Using drums or barrels, small portable SSF units can be set up for household use. Larger plastic tanks can be used for community-scale operations (Figure 5.3). Typically, the filter should be at least 0.5 m deep (ITACA, 2005). The mean diameter of sand typical sand particle size is between 0.15 and 0.35 mm (ITACA, 2005), with a uniformity coefficient of less than 5 (Logsdon et al., 2002). The filters operate most effectively at a continuous and constant flow rate of 0.1-0.3 m/h (or $m^3/h/m^2$) or 100-300 L/h per m^2 of filter media (Logsdon et al., 2002). Generally, smaller sand size and slow flow rates make for the best removal efficiencies (Kubare and Haarhoff, 2010).

For households and small communities, plastic barrels that are 55-gallon capacity (Figure 5.4) can be used to construct the SSF system. These barrels are typically 88 cm tall and 57 cm diameter. In order to study the effectiveness of the removal of particles and other contaminants, it is best to install sampling ports at various depths, starting at a depth of 2 cm below the surface

Figure 5.3 Community-scale slow sand filtration systems using plastic tanks. http://www.daytonsec.org/?p=269 (For the color version of this figure, the reader is referred to the online version of this chapter.)

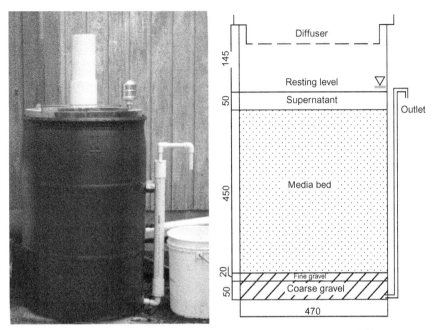

Figure 5.4 A photo (left) and schematic (right) of a single-barrel slow sand filtration unit typically used for household use. Refer to: http://www.shared-source-initiative.com/ biosand_filter/complete_biosand.html for the left photo) (For the color version of this figure, the reader is referred to the online version of this chapter.)

of the sand. A support medium (typically 5 cm of coarse gravel below 2 cm of fine gravel) is used below the sand bed. The underdrain subsystem lies in the gravel. After packing the sieved sand to the desired depth, the sand is saturated with clean water and then back flushed for several hours to remove fine particulates. Then the feed water is supplied to the top of sand. This can be done manually through a siphon mechanism or from an overhead tank. The SSF should have an overflow system to maintain a constant head of water. The filtered water may be additionally treated using silver-impregnated activated carbon filters or UV light (or both).

The authors evaluated the use of SSF for small community water supplies in the Philippines following disasters. In a demonstration study, they used source water from a reservoir in Fort Magsaysay in the Philippines. They used tanks that could be filled manually or with a small pump if power was available to store the source water from the reservoir, and they used standard barrels that are 55-gallon capacity to make the filters. Figure 5.5 shows such a unit near a reservoir.

A key component of the SSF is the *schmutzdecke*, often referred as the biolayer or zoological film. This reddish brown, slimy film acts as a

Figure 5.5 A slow sand filter set up at Fort Magsaysay in the Philippines. (For the color version of this figure, the reader is referred to the online version of this chapter.)

preliminary filter that removes fine colloidal particles and organic matter from the raw feed water (ITACA, 2005). It consists of active bacteria, their wastes and dead cells, and partly assimilated organic matter. Within this film, bacteria derived from the raw water multiply selectively, and the deposited organic matter is metabolized as nutrients. The bacteria oxidize part of the food for energy to grow and reproduce (Huisman and Wood, 1974). Within 2–3 weeks, the *schmutzdecke* is fully developed and ripened, depending on the flow rate, the composition (organic matter present), the oxygen content, and the temperature of the raw water (ITACA, 2005). As water continues to pass through the filter, the *schmutzdecke* thickens, removing contaminants more efficiently. However, once the water passing through the filter slows below the desired flow rate (clogging), the *schmutzdecke* must be scraped and cleaned. After cleaning, it will take a few days to re-ripen (ITACA, 2005).

The quality of feed water affects effluent quality. Turbidity is a key water quality parameter that affects the performance of SSF units. High turbidity water coats the *schmutzdecke* with clay and colloidal particles. This diminishes the biological activity and leads to clogging of the sand surface. Ideally, the turbidity of the feed water should be between 10 and 20 NTU for the SSF to function properly. It can handle higher turbidity for a short time; however, exposure to high turbidity (20-30 NTU or higher) for extended periods will cause clogging and coating of the biolayer, thus reducing filtration effectiveness. Multistage filtration, where a roughing filter precedes an SSF and removes a certain amount of turbidity, can be useful during periods of high turbidity. Cleary (2005) studied the effect of a roughing filter on the removal of microbes during SSF of surface water. Roughing filters (Wegelin, 1983; Wegelin and Schertenleib, 1993) typically consist of gravel particles with diameters ranging from 20 to 4 mm and can be set up in vertical or horizontal flow configurations with variable depths/lengths. Vertical filters can be down-flow or up-flow. Cleary (2005) cites several sources to confirm that using a roughing filter increases the hydraulic loading to SSF (as much as 0.4-0.8 m/h). The primary processes affecting particle removal in roughing filters are gravity settling, interception, and diffusion. Galvis et al. (1996) tested up-flow and horizontal flow roughing filters using source water from a heavily polluted lowland river in Colombia. The turbidity, color, and fecal coliform counts for the source water were 15-1880 NTU, 24-344 true color units, and 7300-396,000 MPN/100 mL, respectively. The filters were 4.3 m long or deep, and the filtration velocity was 0.7 m/h. The removal rates for these parameters were 66.7%, 93.8%, and 95.6% for horizontal roughing filters, respectively. For the up-flow roughing filters, the removal rates for the same parameters were 80%, 97.9%, and 99.4%, respectively. Cleary (2005) set up a multistage filtration system, shown in Figure 5.6. A second configuration (Figure 5.7) was set up for comparison. Turbidity removal for the Pilot System 1 (Figure 5.6) during a period of about 4 months is shown in Figure 5.8. As shown in this figure, the effluent turbidity of the SSF was exceeded only once in response to a rain event at the end of March. Pilot System 2 (Figure 5.7) also performed well (see Figure 5.9) with only one incidence of turbidity reaching 1 NTU.

Because ozone was used as a pretreatment before the roughing filters in Pilot System 1, log removal of coliforms just after ozonation was between 2 and 3 logs. In Pilot System 2, it was less than 1 log after roughing filters, and total removal with the multistage filtration was between 2 and 3 logs.

El-Swaify (2013) studied the performance of household slow sand filters (within plastic barrels as shown in Figure 5.4) in parallel and in series. In parallel, two slow sand filters were subjected to differing flow rates to examine the effect of loading rate on the removal of drinking water contaminants. In series, two filters were run at a given flow rate to examine if the second filter further improved the filtration efficiency.

Figure 5.6 A multistage filtration system, called Pilot System 1, used by Cleary (2005). (For the color version of this figure, the reader is referred to the online version of this chapter.)

Figure 5.7 Pilot System 2, as used by Cleary (2005). (For the color version of this figure, the reader is referred to the online version of this chapter.)

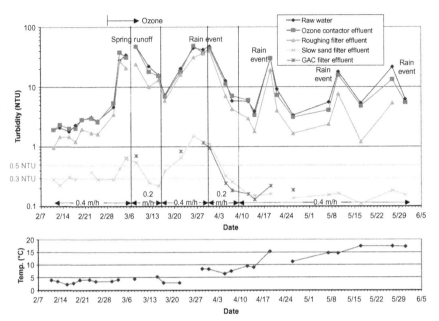

Figure 5.8 Turbidity removal for Pilot System 1 in Cleary (2005). (For the color version of this figure, the reader is referred to the online version of this chapter.)

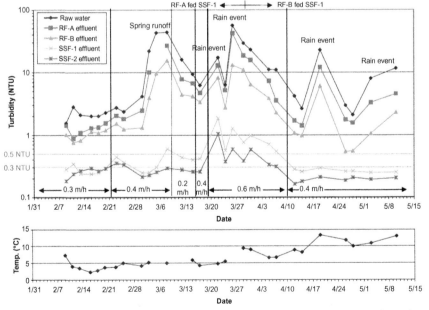

Figure 5.9 Turbidity removal for the Pilot System 2 in Cleary (2005). (For the color version of this figure, the reader is referred to the online version of this chapter.)

In the first parallel mode operation, the water through one filter (B1) had a flow rate of 0.035 m/h, and the water through the other filter (B2) had a flow rate of 0.07 m/h. The filters were run for 1 month. Turbidity, *E. coli*, and total coliforms were monitored in the influent and bottom effluent as well as posttreatment units. The posttreatment units included (a) UV treatment, (b) silver-impregnated activated carbon, and (c) charcoal made from local coconut shells. A Sun-Pure Ust-200 Ultraviolet System (Freshwater System, Greenville, SC) was used for on-line disinfection. The UV bulb was contained within a polypropylene chamber (5 cm internal diameter and 30 cm length) that had a stainless steel lining and a 0.625 cm inlet and outlet. The system allowed a flow rate of 3.6 liters per minute (lpm), with a dose of 30 mJ/cm^2.

A second test was conducted to examine the effect of a sudden change in flowthrough on water quality. El-Swaify (2013) ran water through one filter at 0.025 m/h (B1) and water through another at 0.12 m/h (B2) in parallel operation. He ran those two filters for 14 days and then reversed the flows (high rate to low rate and vice versa [i.e., the new flow rate of B1 became the old flow rate of B2 and vice versa]). Posttreatment units attached to the filter were UV light and silver-impregnated activated carbon.

In series mode, El-Swaify ran the two filters at a flow rate of 0.1 m/h for 38 days. During the first 27 days, the turbidity of the feed water averaged around 5 NTU. After that, the feed water turbidity increased to about 23 NTU.

He also conducted a stress test on the two barrels in parallel configuration (B1 and B2), simulating the effect of a sewage spill in stream water to one barrel. The flow rate was 0.035 m/h (barrel B1), and the simulated spill lasted for 1 day. After that, normal stream water was fed to both the barrels for two more weeks. For barrel B2, no spill was simulated. The flush out effect was studied at the bottom of the barrel and at two sampling depths, 2 cm below the sand surface and 28 cm below.

Silver-impregnated activated carbon may be one possible postfiltration treatment. However, we have already pointed out that such units are not highly effective because the silver gets washed away with time and the carbon surface can become a site for bacterial growth. However, the granular activated carbon (GAC) helps to remove tastes and odors from drinking water.

UV light is another possible posttreatment mechanism. Especially in disaster-affected areas, UV systems can be powered by solar panels or car batteries. A solar panel can charge a deep cycle battery and keep a UV system running for years.

In the parallel mode study conducted by El-Swaify (2013), mentioned above, he used UV light, silver-impregnated GAC, and locally prepared charcoal from coconut shells. For series operation, he only used silver-impregnated GAC. For stress tests, no posttreatment devices were used. Figures 5.10 and 5.11 show the removal of turbidity and *E. coli* for the two flow rates (barrels B1 and B2). There was between 1 and 2 log reductions in *E. coli* after the sand. Barrel 1, which had a lower flow rate, showed better removal of *E. coli* (between 2 and 3 logs); however, the turbidity removal rates were not that different (81% for B1 vs. 78% for B2 in the first 24 days, and 83% and 76%, respectively, in the last 6 days). Logsdon et al. (2002)

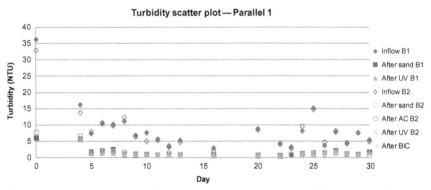

Figure 5.10 Removal of turbidity in two barrel-type slow sand filters; barrel B1 had a flow rate of 0.035 m/h, and barrel B2 had a flow rate of 0.07 m/h. (For the color version of this figure, the reader is referred to the online version of this chapter.)

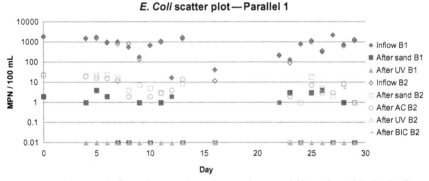

Figure 5.11 Removal of *E. coli* in two barrel-type slow sand filters; barrel B1 had a flow rate of 0.035 m/h, and barrel B2 had a flow rate of 0.07 m/h. (For the color version of this figure, the reader is referred to the online version of this chapter.)

suggest operating at flow rates between 0.1 and 0.3 m/h. In this case, both B1 and B2 had lower flow rates than the minimum suggested.

El-Swaify also found that, at low-flow rates, the biolayer did not fully form during the duration of the study (30 days) whereas, at the high flow rate, it took about 3 weeks for the biolayer to form. Although the formation of the biolayer was not complete in barrel B1, turbidity and bacteria were both steadily reduced. Thus, there must be a sufficient amount of flow and possibly nutrients so that the biolayer can form.

Figures 5.12 and 5.13 show the removal of turbidity and *E. coli* in the two parallel barrels where the flow rates were reversed after 2 weeks. While the removal in turbidity was not significant, even rising in the second half of the

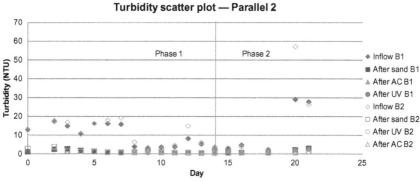

Figure 5.12 Changes in turbidity in two filter units run parallel; the flow rates were reversed in the middle of the experiment. (For the color version of this figure, the reader is referred to the online version of this chapter.)

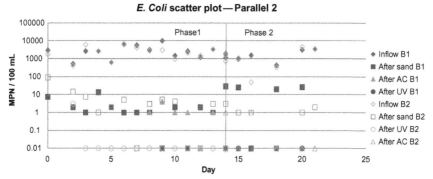

Figure 5.13 Changes in *E. coli* in two filter units run in parallel; the flow rates were reversed in the middle of the experiment. (For the color version of this figure, the reader is referred to the online version of this chapter.)

experiment, the removal of *E. coli* was affected by the changes in flow. For barrel B1, the increased flow rate in the second half of the experiment increased the effluent *E. coli*. Similarly, there was a drop in the *E. coli* for B2 when its flow rate was dropped from 0.07 to 0.035 m/h.

An interesting observation was made when the two barrels were run in series (but at the same flow rate). Overall, a higher removal of turbidity, total coliforms, and *E. coli* was observed in B1 compared to B2 (Figures 5.14 and 5.15).

It can be concluded from this simulation that when the feed water has normal turbidity (3-10 NTU), total coliforms (10^2-10^4 MPN/100 mL), and *E. coli* (10^1-10^3 MPN/100 mL), the first barrel sufficiently purifies the water. Because B1 removed the majority of the "food" (i.e., organic matter, bacteria) from the source water, B2 did not have enough nutrients to develop

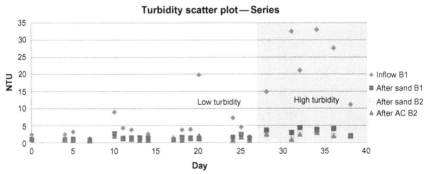

Figure 5.14 Changes in turbidity of effluent when two barrels were run in series. (For the color version of this figure, the reader is referred to the online version of this chapter.)

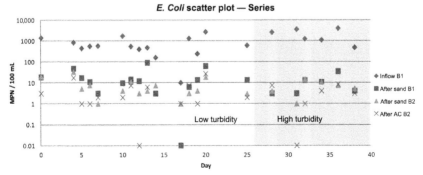

Figure 5.15 Changes in *E. coli* concentration in effluent when two barrels were run in series. (For the color version of this figure, the reader is referred to the online version of this chapter.)

the biolayer. As a result, B1 formed a thick, healthy biolayer, while B2 did not form a biolayer at any time during the experiment. Under normal conditions, one SSF with a GAC column as a posttreatment device is a better option than two SSFs in series.

To improve the efficiency of the two barrels, B1 and B2 can be configured in parallel, maintaining each filter and establishing their respective biolayers. In the event of a stress condition, such as extended periods of heavy rainfall or a sewage spill, B1 and B2 can then be configured in series. Because of the high starting values of total coliforms (10^6-10^7 MPN/100 mL) and E. coli (10^5-10^6 MPN/100 mL) at these times, B1 will most likely shed a large amount of bacteria, which B2 will be able to fully remove. In a case of high turbidity (>15 NTU), a prefiltration technique such as a roughing filter or natural settling should be employed to prevent clogging of the SSF.

The results of stress tests for turbidity and E. coli are shown in Figures 5.16 and 5.17. Figure 5.18 shows the concentrations of total coliforms in water samples collected from the same test. Figure 5.19 shows the removal efficiency for E. coli, coliforms, and turbidity with various posttreatment units.

As shown in these figures, UV disinfection was most effective in reducing the coliforms and E. coli. Activated carbon reduced the turbidity by another 10%; however, its ability to remove coliforms of E. coli was between 50% and 60%. Also, as stated earlier, activated carbon units are not efficient in removing microorganisms.

It is clear that SSF can be viable a treatment technology for long-term applications at varying scales, from household drinking water needs to those of large communities. Additional research characterizing the microbial

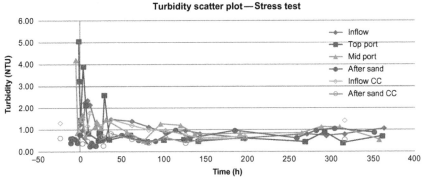

Figure 5.16 Changes in turbidity in the effluent and at various depths in the sand when the filter was subjected to a stress test of high coliform load simulating a spill. (For the color version of this figure, the reader is referred to the online version of this chapter.)

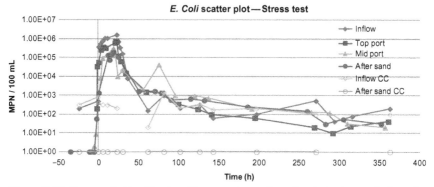

Figure 5.17 Changes in *E. coli* in the effluent and at various depths in the sand when the filter was subjected to a stress test of high coliform load simulating a spill. (For the color version of this figure, the reader is referred to the online version of this chapter.)

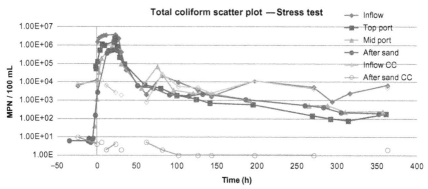

Figure 5.18 Changes in total coliform in the effluent and at various depths in the sand when the filter was subjected to a stress test of high coliform load simulating a spill. (For the color version of this figure, the reader is referred to the online version of this chapter.)

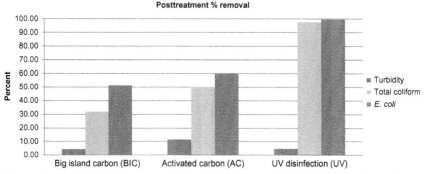

Figure 5.19 Effectiveness of three posttreatment devices in reducing the effluent turbidity, coliforms, and *E. coli*. (For the color version of this figure, the reader is referred to the online version of this chapter.)

community structure and the ability of SSF to remove emerging chemicals such as pharmaceuticals, antibiotics, and endocrine-disruptors will be useful.

5.2 PACKAGED FILTRATION UNITS

Filters that use either micro- or ultrafiltration to purify water have been tested and can be used in emergency water treatment. These packaged filters are powered by either gravity or human suction. Because of this limitation, these filters produce only small volumes of water and are more appropriate for use by small families or individuals. However, because these devices are powered by gravity and suction of human mouth, they are relatively low cost and do not consume fuel so they can be used if no alternative power sources are available. Like most filtration devices, packaged filtration units require proper maintenance and operation to achieve effective bacterial and contaminant removal.

5.2.1 Candle Filter

The candle filter is one of the most researched and evaluated ceramic filters used to purify water. A candle filter is a filter that looks like an upside down candle. A candle filter is screwed into the bottom of the top bucket of a stacked bucket filtration system; see Figures 5.20 and 5.21. Water is poured

Figure 5.20 Clear Kisii filters (Dies, 2003). (For the color version of this figure, the reader is referred to the online version of this chapter.)

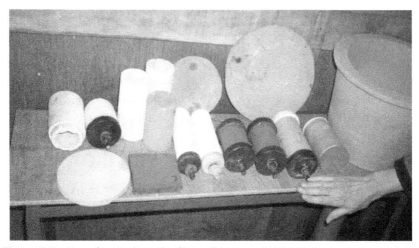

Figure 5.21 Size of typical candle filters (Dies, 2003). (For the color version of this figure, the reader is referred to the online version of this chapter.)

into the top bucket, percolates through the ceramic candle, and is stored in the bottom bucket. The bottom bucket contains a spigot that can be opened for water consumption. The bottom bucket can generally store up to 20 L of water.

5.2.1.1 Materials, Manufacturing, and Removal Efficiency

Ceramic candle filters can be manufactured by professional companies or by locals. Local construction of these filters results in varied quality of the products, which affects the filtration system's effectiveness, while professional products have a higher cost but result in a higher quality product. Local Indian companies make filters by mixing red, white, or black clay with water and sawdust or flour and then compressing it into a mold. The filter is heated in an oven to vaporize the sawdust to create open pores. Pore sizes of the ceramic candle filter range from less than 1 up to 5 μm, depending on the manufacturing techniques and materials used. These pore sizes are small enough to filter out bacteria and protozoa but too large to filter out viruses. However, viruses that have a charge can adsorb to filter surfaces, reducing the number of viruses in the treated effluent. Lab tests performed on locally made filters reveal that disease-causing pathogens are not always removed by candle filters because of cracks that allow water (and microbes) to find "paths of least resistance" and enter the effluent. Flow rates of locally made filters are very slow, ranging from 0.04 to 0.25 L/h.

Filters made by companies such as Katadyn, Hong Phuc, or Kisii (Dies, 2003) have been tested and found to be more efficient than locally made filters for bacterial removal, although they do not remove 99.9% of viruses. The Katadyn product uses a filter media impregnated with fine silver powder and GAC to improve the taste and smell of the drinking water. The Hong Phuc filter uses a diatomaceous earth filter, while the Kisii system offers a choice between a ceramic filter or a filter with GAC and silver. Flow rates for these manufactured systems are faster than the locally produced ceramic filters, ranging from 0.125 to 4 L/h (Dies, 2003).

5.2.1.2 Improve Filter Efficiency

To improve efficiency, different configurations of candle filters can be used. More than one candle filter can be inserted into the stacked bucket system in order to increase the flow rate of treated water. A candle filter manufactured by the Rural Water Development Program (RWD) offers a "jumbo" configuration of a five-candle system to schools and small hospitals.

Coating the filter in colloidal silver has been found to effectively remove microbes. Dies (2003) found that by coating the candle filters in silver, the microorganism log reduction value was increased by 1.15 and the *E. coli* log removal value (LRV) was increased by 0.85.

5.2.1.3 Maintenance

Candle filters require frequent maintenance. Because of particle build up and the resulting reduced flow rates, filters must be cleaned with a soft brush and clean water every 2 weeks (Dies, 2003). After successive cleanings, the ceramic candle tends to lose wall thickness and becomes less effective for removing microbes (Clasen and Boisson, 2006). The service life of candle filters vary, but they should be replaced every 6-12 months (Clasen and Boisson, 2006; Dies, 2003).

Ceramic candle filters are fragile and must be handled with care. Even the slightest crack will reduce the microbiological removal effectiveness of the device. Surveys of candle filter users noted that many people stopped using them because of breakage. Those who owned broken filters rarely bought a replacement filter and went back to using untreated water (Clasen and Boisson, 2006) whether replacement filters were available or not. Training, although not extensive, can promote proper maintenance and use of candle filters. Some shop owners who sell these filters teach consumers how to operate and maintain their new filters (Dies, 2003).

5.2.1.4 Cost
- Indian Company Manufacturers (1-2 candles, containers, spigot): $8-21
  - Purchase of plastic buckets (instead of containers) to save money: $2.22 for buckets
- Nepal Filters: Terracotta clay containers with white kaolin filters for $4.07 each
  - Or buy in bulk for $2.29 each
- Katadyn Filters (containers, spigot, filters): $160-190
- Hong Phuc Filter (containers, spigot, filters): $7.50
- Kisii Filters
  - Slow rate filter: $1
  - High rate filter: $12

5.2.2 Ceramic Disk Filter

The disk filter is another ceramic filter used to purify water. A disk filter is a cylindrical ceramic filter that fits into the bottom of the top bucket of a stacked bucket filtration system; see Figures 5.22 and 5.23. These buckets can be made of terracotta, metal, or plastic. Water is poured into the top bucket, percolates through the ceramic disk, and is stored in the bottom bucket. The bottom bucket contains a spigot that can be opened for water consumption.

5.2.2.1 Materials, Manufacturing, and Removal Efficiency
The Thimi filter is made by locals and the TERAFIL is made by manufacturing companies. The Thimi filter's diameter ranges from 6 to 9 in., and this

Figure 5.22 Terracotta containers to hold disk filters (Dies, 2003). (For the color version of this figure, the reader is referred to the online version of this chapter.)

Figure 5.23 A disk filter (Dies, 2001). (For the color version of this figure, the reader is referred to the online version of this chapter.)

filter is approximately 3 in. thick. A mixture of red clay, ground sawdust, rice husk ash, and water is used to create the desired pore sizes. These materials are all available locally. The materials are mixed by hand, pressed into a mold, and fired in a kiln at certain temperatures for certain times. When the disk is ready, it is cemented into the bottom of a container in a stacked bucket configuration, which can also be terracotta clay. The manufactured disk is similar in construction and operation to the candle filter, except it uses river sand in its clay mixture instead of the rice husk ash and the stacked bucket configuration is metal instead of terracotta clay.

Flow rates for the locally made disks were very low, ranging from 0.1 to 0.3 L/h. The manufactured filter disks had flow rates that ranged from 1.1 to 6.9 L/h. The disks were able to effectively remove turbidity, total coliforms, fecal coliforms, *E. coli*, and iron. Although always greater than 95%, the microbial removal rates were not always to the standard approved by the World Health Organization (WHO). The results varied with different disk filters, even though they were manufactured to have the same specifications, indicating a lack of quality control. Colloidal silver was tested as a microbial disinfectant but, because of varied pore sizes, did not seem to have an effect on microbial concentrations.

5.2.2.2 Maintenance

Most studies on disk filters report that regular cleaning (scrubbing of the filter) is required to maintain filter performance, especially satisfactory flow rates (Dies, 2003; Low, 2001).

The filter can be cleaned with a soft nylon brush to remove accumulated sediment from the top of the disk filter and open new pores. This needs to be done regularly to prevent sediments from slowing the flow rate. Available studies do not mention replacement time for these disks, and the systems did not seem to have easy replacement mechanisms. The seal used to cement the disk into the bottom of the top bucket is a weakness of the design.

5.2.2.3 Cost

- TERAFIL filter (two containers, disk filter): $4.20
- Production cost of the disk filter only: ~$1.00

5.2.3 Ceramic Pot Filters

The pot filter was one of the first ceramic filters manufactured by Potters for Peace. A pot filter consists of a ceramic pot perched inside a larger collection bucket. The inner ceramic pot is usually impregnated with colloidal silver for microbial disinfection; the outer collection pot can be plastic. Water is poured into the inner pot, percolates through the bottom of the pot, and is stored in the collection container (see Figures 5.24 and 5.25). The inner pot is fired as one piece, eliminating the possibility of leakage as is seen with use of both the disk filter and the candle filter.

Figure 5.24 A comparison of untreated water in a pot filter and its treated product (http://www.nature.com/embor/journal/v10/n7/fig_tab/embor2009148_F2.html). (For the color version of this figure, the reader is referred to the online version of this chapter.)

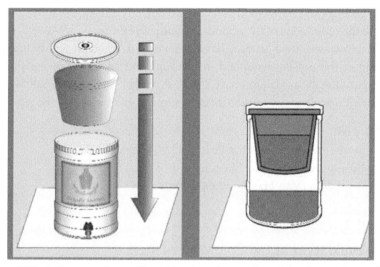

Figure 5.25 A diagram of a pot filter. (For the color version of this figure, the reader is referred to the online version of this chapter.)

5.2.3.1 Materials, Manufacturing, Removal Efficiency

The inner ceramic pot is manufactured similarly to the disk filters. Clay is mixed with water, rice husks, or sawdust and is formed using a simple hydraulic press. After the pot is fired in an inexpensive kiln, it is either soaked in a solution of colloidal silver for thirty seconds or the colloidal silver solution is painted onto the pot using a brush. If silver is added to the collection system, the amount of biofilm growth can be reduced (Murphy et al., 2009). The pore sizes of the pot filter range from 0.6 to 3.0 μm, which will remove protozoa and bacteria although *E. coli* will only be erratically removed. Flow rates are slow and range from 1 to 2 L/h. Salsali et al. (2011) reported that the LRV for *E. coli* was about 2-6 logs and the LRV for protozoa was about 4-6 logs. Removal efficiencies for viruses were less than one log removal.

5.2.3.2 Maintenance

Careful use of the pot filter is required for effective filtration. The inner pot cannot be filled to overflowing because the water could seep through cracks at the very top. According to Resource Development International Cambodia (RDIC), the inside of the pot filter should be cleaned as needed (when the flow decreases substantially), and the storage container should be washed with clean water and soap and air dried in the sun bi-weekly. When performing routine maintenance, the pot filter should not be placed on the ground, hands/tools should be washed before touching any part of the filter apparatus, and the outer edge of the pot filter should not be touched (Murphy et al., 2009). However, field studies have shown that frequent scrubbing and cleaning of the pot have resulted in breakage and failure, reducing the life of the pot filter (Van Halem et al., 2009). A study by Campbell (2005) concluded that pot filters can be used for 2-5 years before they need to be replaced.

5.2.3.3 Cost
- 1 pot filter from Potters for Peace: $9.00 (Dies, 2003)
  - Bulk purchase of pot filters from PFP: $6.00 each
- In Cambodia (Van Halem et al., 2009),
  - Investment costs: $4-8
  - Replacement filters: $2.5-4

5.2.4 Evaluation of Ceramic Water Filters

Ceramic filter systems are low cost and require little maintenance. They can be manufactured locally using local materials. While local manufacturing makes the process more sustainable and helps other people in a developing country make a living, the quality of the products varies, and better quality control is needed. Table 5.2 provides a comparison of the strengths and weakness of using ceramic filters. The average lifetime of these filters is around 1 year, eliminating any long-term applications. According to the Sphere Project (2011), only commercially produced ceramic filters provide the desired flow capacity for emergency survival.

Table 5.2 Strengths and Weaknesses of Ceramic Filters (Dies, 2003)

Strengths	Weaknesses
<ul><li>Relatively cheap to manufacture and produce</li><li>The ceramics trade is well established in many countries</li><li>Materials (clay, sawdust, rice husks) are often readily available</li><li>If designed and used properly, can remove up to 99% of indicator organisms and reduce turbidity to below World Health Organization guideline values.</li></ul>	<ul><li>Very slow filtration rates (typically between 0.5 and 4 L/day)</li><li>Filter maintenance and reliability depends on the user; thus, there are many nontechnical social issues associated with use</li><li>Breakage during distribution or use can be a problem as ceramic filters are often fragile</li><li>Ceramic filters require regular cleaning</li><li>The rate of production (as seen in countries such as Nicaragua and Nepal) tends to be relatively slow</li><li>It is difficult to maintain consistency (quality control is an issue)</li></ul>

5.2.5 Lifestraw Personal

The Lifestraw Personal is a portable water treatment device that operates using human suction. One end of the microfiltration cartridge is placed in dirty water, and suction is applied to the other end (Figure 5.26). The prefilter at the bottom of the cartridge removes coarse materials larger than 1 mm. The vacuum pressure created from the suction drives the water up through the 0.2 μm pore sizes of the microfiltration membrane at an average flow rate of 200 mL/min (Lifestraw, 2008). The membrane achieves 6 LRV for bacteria and 3 LRV for protozoa. Turbidity is also decreased (Lifestraw, 2008). The device can filter approximately 1000 L of water over its lifetime. To clean the device, blow through the suction end to remove dirty particles from the membrane (Figure 5.27).

5.2.5.1 Cost
- The Lifestraw Personal is available in America for $19.95
- The device is available for humanitarian aid and disaster relief for $6 (Peter-Varbanets et al., 2009)

5.2.6 Lifestraw Family

The Lifestraw Family is an ultrafiltration-based system that uses a feed water container elevated above a purification cartridge that contains a hollow fiber membrane (Figure 5.28). The system can last for up to 3 years before needing replacement parts, if properly maintained. Flow rates of 12-15 L/h can

Figure 5.26 Drinking from the Lifestraw. http://www.sswm.info/sites/default/files/reference_attachments/VESTERGAARD%20FRANDSEN%202011%20Life%20Straw.pdf (For the color version of this figure, the reader is referred to the online version of this chapter.)

Figure 5.27 Cleaning the Lifestraw. (For the color version of this figure, the reader is referred to the online version of this chapter.)

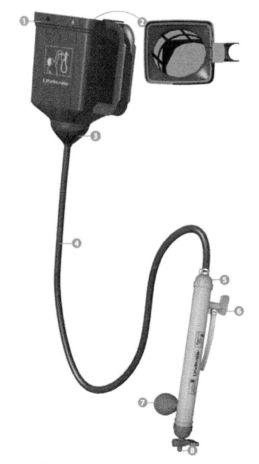

Figure 5.28 Hang the Lifesaver Family to create enough pressure to operate the device. (For the color version of this figure, the reader is referred to the online version of this chapter.)

be achieved with this device. Source waters that can increase the service life of the device include rivers, wells, and rainwater.

The filter consists of a 2 L feed container with a textile prefilter that removes particles larger than 80 μm. Water flows by gravity through the prefilter, and then small amounts of active chlorine are added to the water to keep the membranes from fouling. Water flows down through the 1 m hose to the ultrafiltration membrane, where the 0.1 bar pressure created by the head allows the water to flow through the 20 nm pores and out the tap; after this point, the water should be properly stored. The 20 nm pore size helps the device achieve a removal value of 6 log reductions for bacteria, 4 log reductions for viruses, 3 log removals for protozoan cysts, and helps decrease turbidity (Lifestraw, 2008). Studies have shown stable operation and high efficiency of bacteria and virus reduction during filtration of 18,000 L of source water (Peter-Varbanets et al., 2009).

A manual backwash is included in the system (the red valve shown in Figure 5.29) and should be run daily to remove all particles that clog the membranes. The prefilter used in the feed container should also be washed

Figure 5.29 Configuration of the Lifesaver Family. (For the color version of this figure, the reader is referred to the online version of this chapter.)

daily, while the bucket should be washed weekly. Because of the size constraints of the feed container, constant transport of water and operation of the device is required to provide enough water for consumption.

5.2.6.1 Cost

- This device is not yet available for retail in North America, but the retail price is estimated to be approximately $80 (Amazon.com, 2014).

5.2.7 FilterPen

Similar to the Lifestraw Personal, the FilterPen is a product manufactured by Norit that uses human suction power to pull water through microfiltration membranes with pore sizes of 0.2 μm (Figure 5.30). The small pore sizes

Figure 5.30 The small size of the FilterPen makes it ideal for traveling. https://www. filtertechproducts.com/store/index.php?_a=viewProd&productId=52 (For the color version of this figure, the reader is referred to the online version of this chapter.)

help the device achieve a removal of a 6 log reductions for bacteria, but the device does not achieve effective removal for viruses. The FilterPen is a disposable device that has a service life of about 1 month. The approximate flow rate is 0.1 L/min. After the membranes become clogged, the filter will not allow any flow through the device, indicating that the device should be discarded. Because this device is so small and light (around 30 g), it can be used by travelers, but it can also be used as an immediate, short-term solution in a disaster.

5.2.7.1 Cost
- Prices range from $50-92.50

5.2.8 Chulli (Ovens) Treatment

The chulli treatment method uses thermal waste energy to pasteurize water. Chullis are traditional clay ovens used for cooking in areas of rural Bangladesh. These ovens can reach high internal temperatures, but a large portion of the heat is wasted because of inefficiencies. A water purification system has been designed so a hollow aluminum coil is placed inside the oven. One side of the coil is attached to an elevated untreated water supply source (usually a tank). The tank has a clean sand filtration system to reduce turbidity in the water. The other side of the aluminum coil is attached to a tap that opens and closes to allow water to flow through the coil. See Figure 5.31 for an example of the chulli water treatment setup. When the chulli is used for cooking, the tap can be opened to allow hot water to pass through the aluminum coil. Water is collected in a clean bucket once the water flowing out of the tap becomes too hot to touch safely.

When studied in a laboratory setting, the chulli treatment system had a high removal rate of pathogens. It was also hypothesized that the chulli could provide treatment without extra time or fuel (Islam and Johnston, 2006). However, a field study of chullis already in operation found that around 80% of the operators interviewed stopped using their chulli for water treatment because of mechanical breakdowns or complex and inconvenient operation and maintenance (Gupta et al., 2008). Furthermore, studies showed that the pathogen and coliform reductions were not as high in the field as they were in the laboratory (Gupta et al., 2008). While the ovens were cost effective (chulli owners only paid between $1.50 and $6.00 for the equipment), proper use of the ovens required training. When ovens were used incorrectly, the water treatment

Figure 5.31 Example of a chulli device from Islam and Johnston, (2006). *Credit: Richard Johnston, UNICEF.* (For the color version of this figure, the reader is referred to the online version of this chapter.)

system did not work, discouraging people from using the system. As this system requires more parts and a complicated setup process, it is not applicable in the acute emergency phase. If chullis already exist in a disaster-affected area, using the chulli water purifier system would provide a more sustainable and long-term response.

5.3 PRESSURIZED FILTER UNITS

Pressurized filtration units are devices that have been designed to produce larger volumes of drinkable water than packaged filtration. These devices can be used by small communities or a group of several families. Energy sources can be solar power, hand power, or gravity. These units produce clean water in amounts that range from enough for one person to enough for communities. Some devices can provide high quality water that is suitable for long-term applications, but these devices typically have a high initial investment. However, because the more expensive devices are able to provide drinking water to large communities for a long time, the cost of water per person per day becomes reasonable. Different types of filtration are used, ranging from microfiltration (MF) to reverse osmosis (RO).

5.3.1 Multistage Backpack Filter

This filtration unit combines three processes: A Pentek polypropylene filter is used to remove any suspended solids, a Pentek carbon block filter removes bad tastes and organics, and a hand–cranked UV disinfection unit from Steripen is used to remove viruses, bacteria, and protozoa (Figure 5.32). The entire system is designed so the unit fits within a backpack and can be easily carried from place to place; see Figure 5.33 (Ray et al., 2012). This system is not yet mass produced, but its applicability for humanitarian aid and disaster relief is being assessed by university researchers in Hawaii and by the Thai military (Ray et al., 2012).

Figure 5.32 A backpack filter system. *From Ray et al., 2012,* (For the color version of this figure, the reader is referred to the online version of this chapter.)

A Backpack
B UV disinfection unit
C Sediment filter housing
D Carbon filter housing
E Valve to attach bicycle pump
F Source water bottle
G Valve to attach bicycle pump

Figure 5.33 Filters and UV disinfection device used in a backpack filter system. *From Ray et al., 2012,* (For the color version of this figure, the reader is referred to the online version of this chapter.)

5.3.1.1 Operating Removal Efficiency

The backpack filter system is designed so the unit can be pressurized using a handheld bicycle pump to 7-10 pounds per square inch (psi), or it can operate on gravity using 3 ft of head. The system can efficiently remove coliform bacteria and *E. coli* and can reduce turbidity. The filters do not remove salts or other dissolved solids. The system works as a batch process and can produce 1 L of water in 5 min; thus, it would work best for individuals or small groups of people. Continuous operation of the system is required because the filtered water has to be disinfected with a hand-cranked UV light. This multiphase unit is designed for a temporary and quick response to disasters and has not yet been tested for long-term applications. The unit fits easily within one backpack and can be carried from site to site, reducing waiting time for clean water during a disaster. The final weight of the backpack and system is about 7 kg. Little maintenance is specified because the filters are not expected to be used for long. Filter service life is also not specified, although the hand-cranked UV lamp was specified for 8000 1-L treatments (Ray et al., 2012).

5.3.1.2 Cost
- Pentek spun polypropylene filter: $3
- Pentek carbon block filter: $10
- Steripen UV disinfection unit: $100
- Total: $113
- Cost of production of a full device is unknown

5.3.2 Packaged and Portable RO Filter

The RO system is a six-stage, compact system that is packaged in a Pelican carrying case so it can be easily transported to any fresh source of water (Figure 5.34). The first stage of treatment consists of a polypropylene sediment filter, the second and third stages use carbon block filters to remove odor and organics, the fourth stage uses the RO membrane to remove *Giardia* cysts and *E. coli*, the fifth stage uses a polishing carbon block filter, and the sixth stage uses a 1 L hand-cranked UV disinfection Steripen; see Figure 5.35 (Ray et al., 2012). This system is not yet mass produced, but its applicability for humanitarian aid and disaster relief is being assessed by university researchers in Hawaii and by the Thai military (Ray et al., 2012).

5.3.2.1 Operating Removal Efficiency

This RO filter system is designed to operate with two different power sources for flexibility. The system can be operated with a bicycle pump, or it can be pressurized using power from a car battery, which can be charged

A First stage sediment filter
B Second stage carbon filter
C Third stage carbon filter
D Fourth stage RO filter
E Fifth stage carbon filter
F Sixth stage UV disinfection unit
G Output faucet
H Valve to attach bicycle pump
I Source water bottle

Figure 5.34 Packaged RO unit in a Pelican case. *From Ray et al., 2012,* (For the color version of this figure, the reader is referred to the online version of this chapter.)

with a solar panel. Water flows through the first five stages in a continuous process and then is captured at the outlet with the UV unit. It takes the UV unit 1.5 min to disinfect 1 L of filtered water. Removal efficiencies were very high for coliform, *E. coli*, turbidity, total dissolved solids, and minerals. The system can produce 136-170 L/day, and, if operated continuously for 24 h, can produce 340 L/day. This system is appropriate for a small group or several families. Operation requires supplying the necessary pressure to drive water through the system, batch treating the water with the UV system, and then carefully storing the treated water.

5.3.2.2 Cost
- Osmonics 5-μm rating, Model #1-SED10 polypropylene filter: $9
- KX Extruded 5-μm rating, Model #23-CAB10 carbon block filter: $15
- KX Extruded 5-μm rating, Model #23-CAB10 carbon block filter: $15
- Filmtec 0.0001-μm rating, Model #MEM-45 RO membrane: $65
- Omnipure coconut shell refining carbon 5-μm rating, Model #5-TCR carbon polishing filter: $15
- Steripen UV disinfection unit: $100
- Total: $219
- Cost of production of a full device is unknown

5.3.3 WaterBox
The WaterBox is a packaged solution that uses 10 prefilters and 2 filters that incorporate carbon nanotubes into the media to remove 99.9999% of bacteria, 99.99% of viruses, and 99.9% of cysts from groundwater or fresh

Figure 5.35 Filters used in the packaged RO unit. *From Ray et al., 2012.* (For the color version of this figure, the reader is referred to the online version of this chapter.)

First stage:
Sediment filter cost $9

Second and third stage:
Carbon filter cost $15

Fourth stage:
RO membrane cost $65

Fifth stage:
Carbon filter cost $15

Sixth stage:
UV unit cost $100

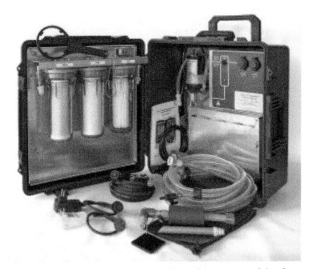

Figure 5.36 The Waterbox (Seldon, 2012). (For the color version of this figure, the reader is referred to the online version of this chapter.)

surface water (Figure 5.36). The WaterBox can also reduce organics, chemicals, and heavy metals found in polluted water. Water is pumped from a source using a manual bicycle pump or power from an AC/DC source or solar panels. The flow rate of the device is around 2 L/min, and the device has a service life of 30,000 L. Because the device is packaged in a case that is about $2' \times 2' \times 1'$, it can be transported directly to water sources and set up within 5 min by an unskilled operator. Hoses and sediment filters must be cleaned periodically to extend the life of the device. Studies were performed on the WaterBox using an independent EPA-certified lab, but no results are available. Because the WaterBox was originally designed for military operations, it may not be available for use in developing countries during emergencies, unless it is donated. This device also produces high quality water rather than a large quantity of water.

5.3.3.1 Cost
- Waterbox sells to military for $7975 (Luquer, 2012)
- Waterbox is offered to humanitarian aid organizations for $3190 (Luquer, 2012)

5.3.4 Lifesaver Jerrycan

The Lifesaver Jerrycan is an ultrafiltration device built into a jerry can. The Jerrycan has an inlet into which water can be poured. The inlet is sealed

Figure 5.37 The Jerrycan has been tested in developing countries. (For the color version of this figure, the reader is referred to the online version of this chapter.)

with a hand pump that is used to pressurize the container. The Jerrycan has a tap at the outlet, preceded by a 15 nm ultrafiltration (UF) filter and an activated carbon filter (Figures 5.37 and 5.38). The 15 nm pore size removes up to 7.5 log reductions of bacteria and 5 log reductions of viruses. Pesticides, heavy metals, and other water pollutants are also reduced. After water is poured into the Jerrycan, the hand pump is used to pressurize the container. The tap can be opened once the container has been pressured enough to release a flow of 2 L/min at 0.1 bar of pressure.

The device can treat up to 15,000 L of water over its lifetime. This can provide a family of four with the minimum amount of water necessary for survival for up to 5 years. The size of the Jerrycan allows storage of up to 18.5 L of treated water. After the filter becomes ineffective, water will not flow through the filter, and the filter should be replaced (Lifesaver, 2011a, b,c). Periodically, the Jerrycan should be rinsed out, and the UF membranes should be unscrewed from the Jerrycan and rinsed in clean water to remove any particulates and to extend the life of the Jerrycan (Lifesaver, 2011a,b,c).

5.3.4.1 Cost
- $270-332 for Lifesaver Jerrycan
- $187-241 for replacement UF filter

5.4 SMALL-SCALE SYSTEMS

Areas such as refugee or internally displaced person (IDP) camps that are set up after emergencies need a large amount of clean drinking water.

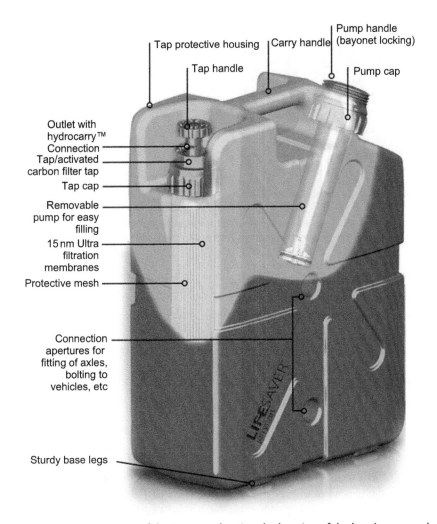

Pump handle
(bayonet locking)

Tap protective housing

Carry handle

Tap handle

Pump cap

Outlet with
hydrocarry™
Connection
Tap/activated
carbon filter tap

Tap cap

Removable
pump for easy
filling

15 nm Ultra
filtration
membranes

Protective mesh

Connection
apertures for
fitting of axles,
bolting to
vehicles, etc

Sturdy base legs

Figure 5.38 Cross-section of the Jerrycan showing the location of the hand pump and ultrafiltration filter. (For the color version of this figure, the reader is referred to the online version of this chapter.)

Small-scale systems can meet these large needs as they usually have the capacity for high flow rates or the potential to be scaled up to achieve high flow rates. High-capacity systems are typically commercially produced devices that provide not only a high quantity of flow but also a high quality effluent. Some systems require little maintenance, while other systems may require a great deal of maintenance. These commercial systems require a large amount of initial capital unless they are donated.

Figure 5.39 Arnal UF system installed in Mozambique (left) and Ecuador (right). (For the color version of this figure, the reader is referred to the online version of this chapter.)

For example, Arnal et al. (2001, 2009) discusses a UF system developed for third-world communities (Figure 5.39). This system can use either generators or a manual rotation wheel that produces energy. The capacity of the system can also be increased by adding more ultrafiltration modules. Veolia produces compact skid-mounted systems, not pictured here, that use ultrafiltration membranes; these systems are used during emergencies (Peter-Varbanets et al., 2009).

5.4.1 Sunspring

The Sunspring device is a self-contained, rugged UF unit that can treat groundwater it pumps from wells or from surface water sources. Seventeen of these devices were implemented during the Haiti disaster relief effort and are still supplying water to the communities of Haiti.

The treatment system is made up of a 4.5 gpm pump, a prefilter, a UF system from GE, plumbing, and an automatic backwash that removes impurities (Figure 5.40). The filters are encased within an aluminum cylinder that provides protection. Two solar panels provide power during the day and charge internal batteries, which provide power to treat water during the night. The device is certified by the WQA Gold Seal Program, which means that it removes 99.999% of pathogens from any source water. It can treat 2000-5000 gallons per day, depending on source water conditions and the amount of sunlight. This quantity of water can provide 10,000 people the minimum amount of water needed to survive (2 L/day). The system can be installed within 2-4 h and used almost immediately. The device

Figure 5.40 The Sunspring uses solar panels to power its pump and treat the water. (For the color version of this figure, the reader is referred to the online version of this chapter.)

can last up to 10 years with minimal maintenance. The membranes should not be allowed to dry out or freeze.

The Sunspring comes with a maintenance kit for the required minimal maintenance. Prefilters should be changed weekly to quarterly, depending on the source water quality. Ensuring that the prefilters are clean will help keep the membrane modules clean. Membrane modules need to be changed every 10 years, also depending on the source water quality. In order to test that the Sunspring is effectively removing all pathogens and viruses, membrane integrity testing (MIT) needs to be performed periodically. MIT kits also come with the Sunspring.

5.4.1.1 Cost

The price per Sunspring varies, depending on what is included inside the casing. In the Haiti system, attachments were added to make the system flexible. These attachments included a surface pump for surface water, a submersible pump for wells 30 m deep, and at least a year's supply of prefilter cartridges. This system costs $25,000. At this price, and assuming the system provides enough water for 10,000 people per day for 10 years, the cost per person per day of these units drops to less than 0.01 dollars per person per day.

- Investment cost: ~$25,000
- Cost does not include maintenance

5.4.2 Perfector-E

The Perfector-E water purification system from X-Flow (NORIT, the Netherlands) was designed to supply water to tsunami victims in Asia by treating heavily polluted surface water (Figure 5.41). According to manufacturer information (Pentair, 2012), it uses a two-stage pretreatment process consisting of a coarse filter and two parallel microstrainers. The main treatment stage consists of two UF dead-end modules operated with automatic backflushing as well as an optional UV disinfection barrier (Peter-Varbanets et al., 2009). The lifespan of the UF modules ranges from 5 to 7 years, depending on the source water quality. Then the modules should be replaced. The only maintenance that is required is regular cleaning.

The flow rate of this device is around 2000 L/h, and it is powered by a generator that comes with the unit. Perfector-E is able to achieve 99.9999% of bacteria removal and 99.99% of virus removal, even in highly turbid waters. Once set up, the system may be operated by unskilled personnel, and the rugged design allows it to be transported from place to place (PWN Technologies, 2012).

Figure 5.41 The Perfector-E uses X-Flow technology. (For the color version of this figure, the reader is referred to the online version of this chapter.)

5.4.2.1 Cost

• Approx. $ 26,000 (Peter-Varbanets et al., 2009)

5.4.3 SkyHydrant

The SkyHydrant is a filtration device that uses UF membrane filtration to treat water for a community. This system can be used individually or in series to increase capacity. One SkyHydrant can produce 500-700 L/h at an operating pressure of approximately 5.7 psi. The SkyHydrant can be used in conjunction with a raw water supply tank that provides the pressure required to power the SkyHydrant. The treated water can be stored in a clean storage tank to provide a continuous supply of treated water; see Figure 5.42 for a drawing of the layout. Communities that already have rooftop water tanks can integrate the SkyHydrant into their piping system (Figure 5.43). This system is complicated to set up as it requires piping, a pump, and storage tanks (Figure 5.43). Thus, the SkyHydrant may be more useful as a long-term, sustained response to a disaster rather than as a short-term solution. The SkyHydrant removes bacteria, viruses, protozoans, and turbidity from source water. Daily cleaning is required to keep the membrane working efficiently although the dirtier the water, the more often the filter will need to be cleaned. According to the operation manual (Skyjuice Foundation, 2010), the UF module does not need to be replaced as long as it is cleaned

Figure 5.42 The SkyHydrant setup requires a raw water pump, a raw water supply tank that feeds to the SkyHydrant, and a drinking water storage tank (Skyjuice Foundation, 2010). (For the color version of this figure, the reader is referred to the online version of this chapter.)

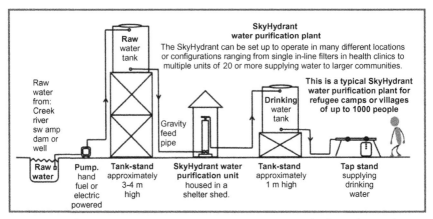

Figure 5.43 The SkyHydrant can be used in series to increase the flow of treated water (Skyjuice Foundation, 2010). (For the color version of this figure, the reader is referred to the online version of this chapter.)

thoroughly daily. Use of the SkyHydrant is complex so a skilled operator should be designated to manually backwash and clean this system with chemicals periodically (Skyjuice Foundation, 2010).

5.4.3.1 Cost
- SkyHydrant: $1000–2000 per unit
- Does not include costs for infrastructure: a raw water storage tank, a treated water storage tank, and a raw water pump

5.4.4 iWater Cycle

In areas where electricity is not available, a bicycle-powered UF membrane system can be used. iWater Cycle has a capacity of 400–900 L/h and can remove turbidity and microbiological contaminants to World Health Organization standards (Figure 5.44). The device is mobile and easy to maintain and operate (Idro, 2010). Maintenance includes manual backwashing when operation pressures exceed normal. This device has been used in Yemen, Myanmar, and Taiwan for emergency relief efforts after the acute emergency phase (Figure 5.45).

5.4.4.1 Cost
- iWater Cycle: $3000

Figure 5.44 iWater Cycle uses a bicycle to power pumps. (For the color version of this figure, the reader is referred to the online version of this chapter.)

Figure 5.45 Victims of a typhoon in Taiwan use bicycle power to filter their water. (For the color version of this figure, the reader is referred to the online version of this chapter.)

5.4.5 Evaluation of Small-Scale Systems

Peter-Varbanets et al. (2009)noted that, "The investment costs of small scale systems are generally too high for communities in developing countries.... These communities also lack trained personnel to maintain these systems and often do not want or cannot assume the responsibility for their performance after their construction by the government or NGOs. The provision of regular maintenance by regional maintenance centers, such as is practiced in some regions in South Africa, needs a certain organization and control and would not be possible in many other countries."

5.5 NATURAL FILTRATION

Natural filtration is a generic term applied to riverbank, lake bank, or SSF. Additionally, desalination of water derived from seashores (beach wells) is considered a natural filtration process. SSF has been described in detail earlier. Here, we will focus on riverbank filtration with some preliminary introduction to lake bank filtration and beach wells for desalination. We will use the generic name "bank filtration" (BF) to represent this natural filtration process.

Bank filtration is a mechanism by which communities located along rivers or lakes develop water supplies from alluvial aquifers using vertical and horizontal (collector type) wells. The sand and gravel deposits comprising these alluvial aquifers yield millions of gallons of fresh water to these communities. When wells are placed sufficiently close to rivers and pumped, river water can be induced to flow to these wells. Bank filtration wells typically produce a larger quantity of water than similarly sized wells that only draw groundwater because the river acts as a steady source. Thus, bank filtration wells have been used by many riverbank communities to augment the groundwater yield that can be expected from an aquifer. The portion of riverbank filtrate in the pumped raw water depends on source water quality, geohydrologic conditions of the aquifer, river-aquifer interface, hydraulic gradient, infiltration rates, hydraulic conductivity, and distance between the riverbank and the pumping wells.

The use of bank filtration for drinkable water dates back more than 100 years. In the lower Rhine region of Germany, riverbank filtration systems have been operating since the 1870s (Schubert, 2002). In the eastern part of Germany, waterworks for the city of Dresden have also operated for more than 100 years (Saloppe since 1875; Tolkewitz since 1898; Hosterwitz since 1908; see Fischer et al., 2006; Ray and Jain, 2011). In the United States, this technology was used initially for industrial water supplies and then for public supplies. The city of Peoria, Illinois, has used bank filtration systems to augment its water supply since the 1940s (Horberg et al., 1950; Schicht, 1965; Marino and Schicht, 1969). Because bank filtration has been shown to be a long-lasting filtration technology, it can be adapted to provide water to affected communities with minimal investment.

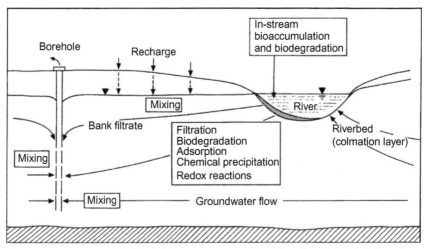

Figure 5.46 Diagram of a bank filtration system. *After Hiscock and Grischek (2002).*

Figure 5.46 shows a diagram of a bank filtration system where a well is located on the bank and the pumping action brings river water to the well. Water from the land side also contributes to the flow to the well. A number of physical, chemical, and biological processes occur as the water moves from the river to the well. Particles that contribute to turbidity are strained out or removed by colloidal filtration (Yao et al., 1971). The temperature of the river water is modulated during its passage through the aquifer. Mixing with groundwater also occurs. Chemical processes such as sorption, ion exchange, biodegradation, geochemical dissolution, and chemical precipitation can occur, depending on the water chemistry and minerals present in the aquifer. Removal of protozoa may occur via straining or colloidal filtration. Bacteria and viruses are primarily removed through colloidal filtration. The redox status of the subsurface affects these chemical and biological processes.

The wells can be either vertical wells or horizontal collector wells. Vertical wells typically have low production rates compared to horizontal collector wells. The preference for a horizontal collector well rather than vertical wells depends on several factors: site hydrogeology, land ownership near the riverbank, and utility preference. In the United States, the preference has been for horizontal collector wells because a large quantity of water can be extracted from one or a few wells. In temperate climates, using a horizontal collector well also reduces the maintenance needs for the well's pumps and pipes during the winter months as multiple vertical wells will

Figure 5.47 Vertical wells at the Flehe waterworks. A monitoring well, which is used to detect chemicals and pathogens in river water in the event of spills or floods, is located between the river and the vertical wells. (For the color version of this figure, the reader is referred to the online version of this chapter.)

be needed to produce the same amount of water. If the source water is high quality, water produced by BF may not need additional treatment except for chlorination or some other method of disinfection. Horizontal collector wells are often used for pretreatment in the United States. The filtrate passes through the treatment plant; however, chemicals are rarely used for coagulation as the water already has low turbidity.

For emergency water needs, construction of horizontal collector wells may be too expensive. However, vertical wells can be easily sunk along the banks of the rivers or streams. Figure 5.47 shows a series of wells at the Flehe Water Works in Dusseldorf, Germany. These wells are connected by underground siphon systems to a central pumping system. For small systems, wells can be equipped with vertical turbine or submersible pumps.

5.5.1 Design of Wells

Wells should be designed so that they pump safe and sand-free water for a long time (at least a decade). In unconsolidated formations, gravel may be used in the annular space between the well screens and the formation, depending on the grain size distribution of the natural formation. In some cases, gravel may not be needed if the formation has grains in the appropriate range. Bedrock wells are different from screened wells in an unconsolidated formation. Most bedrock wells are open holes. The diameter of the well has little influence on the specific capacity of the well (Table 5.3). Specific capacity is defined as the well yield per unit of drawdown (Q/s) and is measured in units of gpm per ft of drawdown. Figure 5.48 shows the variation in specific capacity as a function of well radius. Typically, the casing of the well

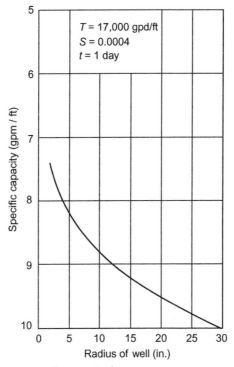

Figure 5.48 Changes in specific capacity.*

Table 5.3 Recommended Well Diameters (Driscoll, 1986)

Expected Yield (gpm)	Pump Bowl Diameter (in.)	Casing Internal Diameter (in.)	Smallest Size of Casing (in.)
<100	4	6	5
75–175	5	8	6
150–400	6	10	8
350–650	8	12	10
600–900	10	14	12
850–1300	12	16	14
1200–1800	14	20	16
1600–3000	16	24	20

must be at least 2 in. greater than the pump bowl diameter. Two types of pumps can be used in the wells: submersible and vertical turbine pumps.

These pumps are used within wells if the water table is deeper than the suction lift of centrifugal pumps. Typically, centrifugal pumps are not

---

*"Groundwater Workshop Participant Handbook". Joint Workshop by the Illinois State Water Survey and Illinois Section of the American Water Works Association. Ramada Inn Convention Center, Champaign, IL. Held February 21–22, 1983.

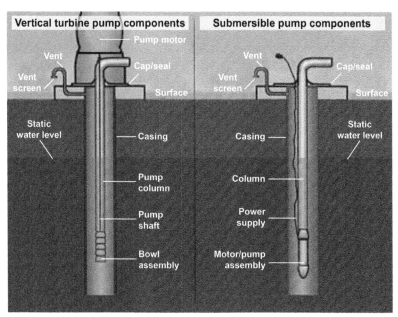

Figure 5.49 Diagram of wells with vertical turbine and submersible pumps (Florida Department of Environmental Protection, http://www.dep.state.fl.us/central/home/DrinkingWater/FieldCompliance/PhysicalComponents/PhysicalComponments.htm). (For the color version of this figure, the reader is referred to the online version of this chapter.)

suitable for wells. Figure 5.49 shows two wells: one with a vertical turbine pump (left) and the other with a submersible pump (right). In the case of a turbine pump, the pump bowl stays in the well (typically above the screen zone), and the motor sits on a pad above the wellhead. The motors for high-capacity wells can be very large (several hundred horse power). Submersible pumps have sealed motor and pump units that are inserted into the well. The power supply cable connects from a power source to the pump. As the diameter of the motor is limited by the diameter of the well, the production capacity of wells equipped with submersible pumps is expected to be less than those of wells equipped with vertical turbine pumps.

The wells are vented to the atmosphere so that negative pressures do not develop as drawdown occurs in the well. Vent screens prevent foreign objects from entering the wells. The wells need to have a cap or seal to prevent surface water or other contaminants from directly entering the well (see Figure 5.49). Wellheads with turbine and submersible pumps are shown in Figure 5.50. A diagram of a sanitary seal is shown in Figure 5.51.

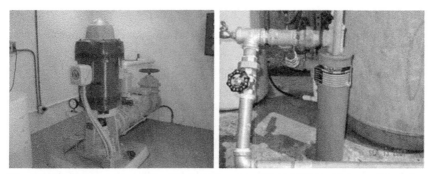

Figure 5.50 Motor mount of a turbine pump on a well head (left), and the outlet of pipe from a submersible pump inside a well (right) (Florida Department of Environmental Protection, http://www.dep.state.fl.us/central/home/DrinkingWater/FieldCompliance/PhysicalComponents/PhysicalComponments.htm). (For the color version of this figure, the reader is referred to the online version of this chapter.)

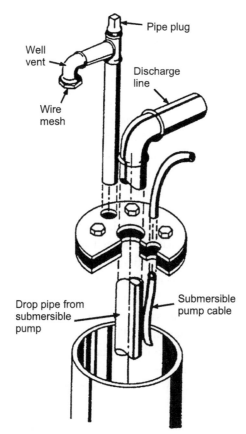

Figure 5.51 Diagram of a sanitary seal for a well head with a submersible pump inside (Florida Department of Environmental Protection, http://www.dep.state.fl.us/central/home/DrinkingWater/FieldCompliance/PhysicalComponents/PhysicalComponments.htm).

Wells constructed in unconsolidated (sand and gravel) formations can be either tubular or gravel packed. In a tubular design, the well screen is placed directly against the sand and gravel aquifer. The screen opening is chosen so that it can retain a certain percentage of the aquifer material. Well development, a technique in which the fine particulates from around the screen are removed by water or air or surging[1], is conducted to enhance the production capacity of the well and to reduce the potential for clogging. In tubular well design, careful well development is necessary to remove the fine particulates.

A significant factor in the design of tubular wells is the size of the screen opening (or slot size). The optimum slot size depends on the size distribution of the aquifer material that will stay in contact with the screen. Two important terms are used to define the size distribution of the material: (a) uniformity coefficient (C_u) and (b) effective size (D_{10}). Uniformity coefficient is defined as:

$$C_u = \frac{40\% \text{ retained size}}{90\% \text{ retained size}}$$

Here "retained size" means the diameter at which a given percentage of the material is retained above the screen. Effective size is 90% of the retained size. A sieve analysis is performed on the aquifer material to obtain this information where the aquifer material is loaded to the top sieve in a nest of sieves of different sizes. Figures 5.52 and 5.53 show sieve analysis data for two different materials—one more homogeneous and the other more heterogeneous. As observed from these figures, the uniformity coefficient for the homogenous material is 2.2 and that for the heterogeneous material is 6. Thus, a homogeneous material will have a size distribution in a narrow band compared to a heterogeneous material. Screen slot size is based on the homogeneity of the aquifer material as well as the stability of the overlying deposits. When the uniformity coefficient is between 3 and 6 (relatively uniform; see Figure 5.52), a screen is selected that retains 40% of the aquifer material if the overlying material is firm (e.g., clay). If the overlying material is subject to caving (such as fine sand), 60% of the aquifer material surrounding the screen must be retained. For heterogeneous aquifer material ($C_u \geq 6$), 30% of the material should be retained for a clay-type overburden, or 50% of the material should be retained if the overlying material is subject to caving. In field situations, conditions often do not fit into these categories,

[1]Surging: It is a reciprocating action in which a plate that is slightly lower than the internal diameter of the well is moved up and down (typically by the drilling rig). This action is called surging and it dislodges fines and brings them to the well.

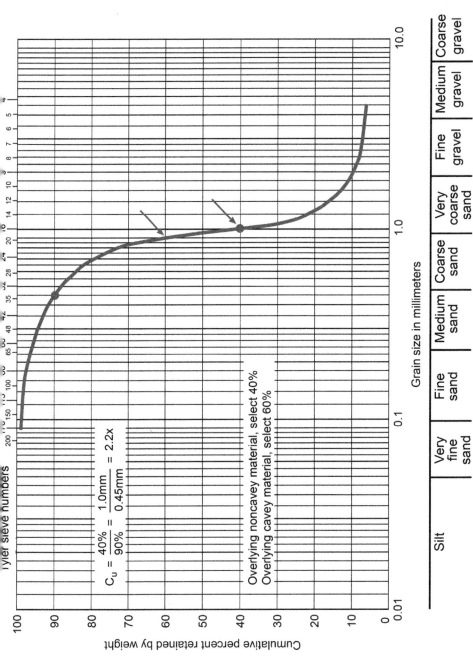

Figure 5.52 Sieve analysis data for a homogeneous medium.

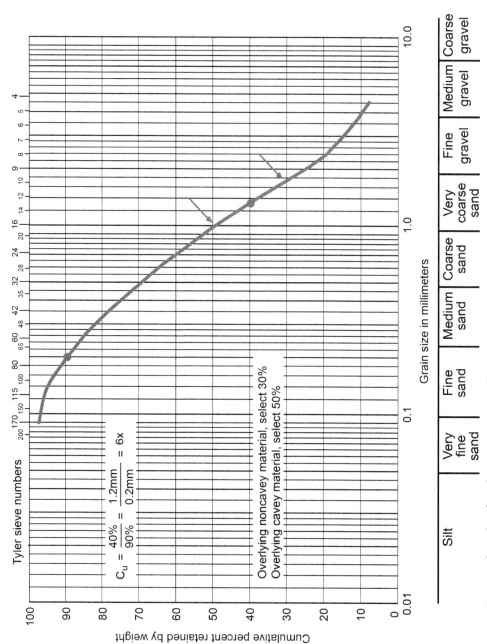

Figure 5.53 Sieve analysis data for a heterogeneous medium.

thus necessitating some judgment. For example, a screen that retains 40% of the material in a heterogeneous aquifer may be selected if the overlying material is sandy clay with few caving tendencies.

As the aquifer becomes more homogeneous, slot size selection becomes more critical. Small changes in the screen opening result in large variances in the amount of retained material for a homogeneous aquifer because of the steep slope of the particle size distribution graph, as shown in Figure 5.52.

If the aquifer has distinct layers with differing grain size distributions, the slot size of the screens may be tailored to the individual layers. In this case, the 50% size for the finest and the coarsest layers are compared. If the 50% size of the coarsest layer is at least 4 × larger than that of the finest layer (see Figure 5.54), then the screen size should be based on individual layers. If the ratio is less than 4:1, then the screen size should be based on the finest interval in the aquifer. Screen tailoring is not necessary if the total screen length is less than 10 ft. Often, the capacity gained by screen tailoring will not be significant as the well log information may not be sufficiently accurate to warrant changes over short lengths.

The effective open area of the screen (A_e) is determined using casing diameter and slot size (determined from above) as well as the screen manufacturer's tables for telescope well screens (see Table 5.4 for data about stainless steel screens and Table 5.5 for data about PVC screens). Because the sand or gravel sits next to the screen openings, a blockage factor of 50% is used when determining A_e.

The screen entrance velocity (V_e) is empirically related to aquifer hydraulic conductivity, as shown in Table 5.6. Scientists at the Illinois State Water Survey developed this relationship based on actual case histories of well failures caused by partial clogging of the well walls and screen openings. Unless hydraulic conductivity data exist for a nearby well, one has to estimate the aquifer property. For example, in Illinois, the hydraulic conductivity values for most sand and gravel aquifers range from 700 to 2000 gpd/ft^2. From specific capacity data, one can estimate the hydraulic conductivity of the aquifer by conducting a pumping test.

Pipe schedules are also standard in North America. Table 5.7 shows the inner diameter, outer diameter, wall thickness, and nominal weights (lb/ft of pipe length) for schedule 40 and 80 pipes. For example, for a 140-in. pipe, the wall thicknesses for 40 and 80 schedule pipes are 0.437 and 0.75 in., respectively.

If the hydraulic conductivity data for the wells from pump tests show that a given velocity should be used, but the nearby wells have experienced clogging or chemical encrustation, then a lower entrance velocity may be chosen.

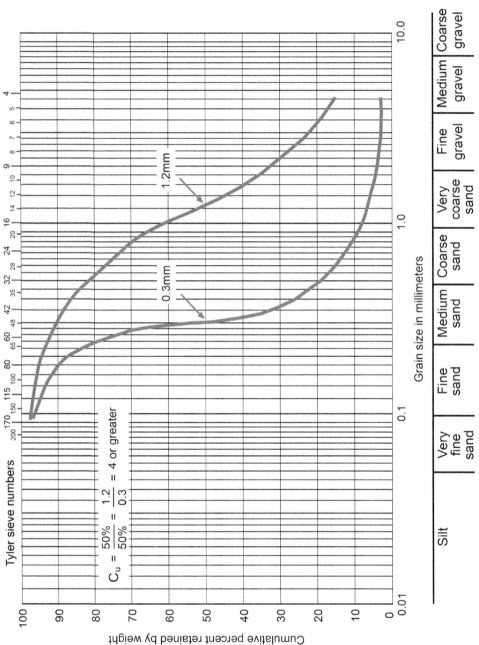

Figure 5.54 The use of 50% size ratios between two different layers for screen tailoring.

Table 5.4 Screen Open Areas for Stainless Steel Screens of Different Slot Sizes Typically Used in North America, for Various Diameters of Pipes Large Diameter Free-Flow Screens: Sizes 6P-16T (from Johnson Screens)

Size (in.)	Max Depth (ft)	OD (in.)	ID (in.)	Weight[1] (lb/ft)	Recom. Hang Weight[2] (lb)	Collapse Strength[1] (PSI)	Intake Area[3] (sq in./ft of Screen) Screen Slot Size (Thousandths of an Inch)							
							10	20	30	40	50	60	80	100
6P	100	6.5	6.0	4.4	4300	87	35	61	82	98	111	123	140	153
	250	6.6	6.0	4.8	4300	194	20	37	51	64	75	85	102	115
	600	6.7	5.9	6.0	8800	185	20	37	52	65	76	86	103	117
	1000	6.8	5.9	7.6	8800	677	16	30	43	54	64	73	89	103
8T	250	7.6	6.7	7.0	11,000	127	23	42	59	73	86	98	117	133
	1000	7.7	6.7	8.9	11,000	468	18	34	48	61	73	83	101	116
8P	250	8.7	7.9	7.9	12,100	85	26	48	67	84	99	112	134	152
	1000	8.8	7.9	10.1	20,800	314	21	39	55	70	83	95	115	133
10T	250	9.5	8.6	8.3	12,100	65	28	53	74	92	108	122	146	166
	1000	9.6	8.6	10.7	12,100	242	23	43	60	76	90	103	126	145
10P	500	10.8	9.8	12.6	15,400	170	25	48	68	86	102	116	142	163
	1000	10.8	9.8	17.8	15,400	226	25	48	68	86	102	116	142	163
12T	600	11.4	10.4	13.6	17,600	145	27	51	72	90	107	123	149	172
	1000	11.4	10.4	19.0	17,600	192	27	51	72	90	107	123	149	172
12P	250	12.8	11.8	14.8	17,600	103	30	57	80	102	121	138	168	193
	600	12.8	11.8	20.9	17,600	136	30	57	80	102	121	138	168	193
	1000	12.9	11.8	25.2	17,600	193	29	55	78	98	117	134	163	188
14T	250	12.6	11.6	13.6	14,300	108	30	56	79	100	119	136	165	190
	600	12.6	11.6	19.6	14,300	143	30	56	79	100	119	136	165	190
	1000	12.6	11.6	24.0	14,300	207	28	53	76	96	114	131	160	184
14P/ 16T	250	14.1	13.1	15.5	17,100	77	33	63	89	112	133	152	185	213
	600	14.1	13.1	22.2	17,100	102	33	63	89	112	133	152	185	213
	1000	14.1	13.1	27.2	17,100	148	32	60	85	107	128	146	179	206

[1]Based on 0.030 inch slot size (collapse values contain no safety factor)
[2]Recommended hang weight is 50 percent of the calculated tensile strength
[3]Transmitting capacity in gpm/ft of screen = open area × 0.31 at 0.1 ft/sec

Table 5.5 Screen Open Areas for PVC Screens of Different Slot Sizes Typically Used in North America, for Various Diameters of Pipes
PVC Schedule 40 Screen Open Area: STD Construction (sq in./ft) (from Johnson Screens)

Pipe Size (in.)	Slot Spacing (in.)	Standard Slot Opening (in.)									
		0.010	0.015	0.020	0.025	0.030	0.040	0.050	0.060	0.100	0.125
½	3/16	0.76	1.11	1.45							
¾	3/16	0.84	1.22	1.59							
1	3/16	1.14	1.67	2.17	2.65	3.10	3.96	4.74	5.45	7.83	9.00
1-1/4	3/16	1.71	2.50	3.25	3.97	4.66	5.93	7.11	8.18	11.74	13.50
1-1/2	3/16	2.05	3.00	3.90	4.24	4.97	6.33	7.58	8.73	12.52	14.40
2	3/16	2.51	4.00	5.20	6.35	7.45	9.49	11.37	13.09	18.78	21.60
2 Hi Flow	1/8	3.56	5.14	6.62	8.00	9.29					
2-1/2	3/16	2.89	4.22	5.49	6.00	7.03	8.97	10.74	12.36	17.74	20.40
3	3/16	3.19	4.67	6.07	8.47	9.93	12.66	15.16	17.45	25.04	28.80
4	1/4	3.12	4.58	6.33	7.77	9.16	11.79	14.25	16.55	24.43	28.50
4 Hi Flow	1/8	6.00	8.68	11.17	13.50	15.68					
5	1/4	3.29	4.84	7.00	8.59	10.13	13.03	15.75	18.29	27.00	31.50
6	1/4	3.23	8.15	10.67	13.09	15.43	19.86	24.00	27.87	41.14	48.00
8	1/4	4.33	9.21	12.06	14.80	17.44	22.45	27.13	31.50	46.50	54.25
10	1/4			14.22	17.45	20.57	26.48	32.00	37.16	54.86	64.00
12	1/4			18.33	22.50	26.52	34.14	41.25	47.90	70.71	82.50
14	1/4							50.00	58.05	85.71	100.00
16	1/4							51.25	59.50	87.86	102.50

Table 5.6 Optimum Screen Entrance Velocities for Various Hydraulic Conductivity Values of an Aquifer

Hydraulic Conductivity (gpd/ft^2)	Optimum Screen Entrance Velocity (ft/min)
>6000	12
6000	11
5000	10
4000	9
3000	8
2500	7
2000	6
1500	5
1000	4
500	3
<500	2

Table 5.7 Pipe Schedules: Especially Schedules 40 and 80 for Different Diameters

Schedule 40 and 80		Schedule 40			Schedule 80		
Pipe Size (in.)	OD (in.)	Avg ID (in.)	Min Wall (in.)	Nom Wt (lb/ft)	Avg ID (in.)	Min Wall (in.)	Nom Wt (lb/ft)
0.50	0.840	0.608	0.109	0.161	0.528	0.147	0.202
0.75	1.050	0.810	0.113	0.214	0.724	0.154	0.273
1.00	1.315	1.033	0.133	0.315	0.935	0.179	0.402
1.25	1.660	1.364	0.140	0.426	1.256	0.191	0.554
1.50	1.900	1.592	0.145	0.509	1.476	0.200	0.673
2.00	2.375	2.049	0.154	0.682	1.913	0.218	0.932
2.50	2.875	2.445	0.203	1.076	2.289	0.276	1.419
3.00	3.500	3.042	0.216	1.409	2.864	0.300	1.903
4.00	4.500	4.998	0.237	2.006	3.786	0.337	2.782
5.00	5.563	5.017	0.258	2.726	4.767	0.375	3.867
6.00	6.625	6.031	0.280	3.535	5.709	0.432	5.313
8.00	8.625	7.943	0.322	5.305	7.565	0.500	8.058
10.00	10.750	9.976	0.365	7.532	9.492	0.593	11.956
12.00	12.750	11.890	0.406	9.949	11.294	0.687	16.437
14.00	14.000	13.072	0.437	11.810	12.410	0.750	19.790
16.00	16.000	14.940	0.500	15.416	14.214	0.843	25.430

The length of the well screen is determined by the following equation:

$$L_s = \frac{Q}{7.48 A_e V_c}$$

where the pumping rate is in gallons per minute, A_e is in ft^2/ft of pipe, V_c is the critical entrance velocity in ft/min, and L_s is screen length in ft. Once the screen length has been determined, one should check if this screen length is both realistic and physically feasible. For example, if a 30 ft screen length is needed but only 25 ft of aquifer exist, then the screen length poses a problem. When finer materials exist above or below the aquifer, the screen should be placed at least 1 or 2 ft from the boundary of the aquifer joining these low-permeability layers. A typical design for a tubular screen well is shown in Figure 5.55.

After determining the depth of the top of the well screen, the available drawdown (s_a) can be determined. The nonpumping water table (found from other wells or measured) is subtracted from the depth of the top of the screen to find s_a. Do not allow the formation to dry out to the screen zone as aeration can promote metal precipitation and encrustation. From the specific capacity data for the well, the maximum pumping rate can be determined. The specific capacity is estimated by comparing the Q/s of nearby wells. A theoretical specific capacity for a new well can be estimated from the following equation:

$$Q/s = \frac{T}{264 \log\left[\dfrac{Tt}{2693(r_w)^2 S}\right] - 66.147}$$

where: $Q/s =$ specific capacity, in gpm/ft; $Q =$ discharge, in gpm; $s =$ drawdown, in ft; $T =$ transmissivity, in gpd/ft; $S =$ coefficient of storage; $r_w =$ nominal radius of the well, in ft; $t =$ time after pumping began, in min. Because conditions rarely lend themselves to just one design option, an alternative design should be evaluated for comparison. This is especially important in cases in which both a natural backfill and artificial (gravel) pack are possible. While in the past both uniform and graded gravel packs were used, these days only uniform gravel packs are used. In general, a gravel pack is needed if the effective size of the aquifer material is less than 0.01 in. and C_u is less than 3.

The ideal size of the gravel pack is based on the 50% size of the finest interval of sand and gravel that needs to be retained by the pack. This includes units of sand and gravel or other unstable formations that the gravel pack contacts,

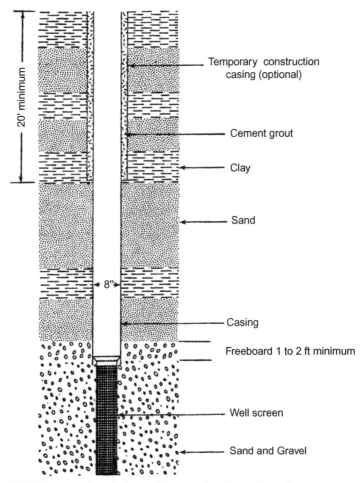

20' minimum

Temporary construction
casing (optional)

Cement grout

Clay

Sand

8"

Casing

Freeboard 1 to 2 ft minimum

Well screen

Sand and Gravel

Figure 5.55 Diagram of a tubular well in a sand and gravel aquifer.

even though they are at a level at the top of the screen. If the exact sizes of these materials are not known, but there are indications that they are fine grained, care should be taken to segregate these fine materials from the gravel pack.

The 50% size of the formation is multiplied by 3 to obtain the 95% retained size of the pack and multiplied by 5 to obtain the 5% retained size. The particle size data for a commercially available gravel pack material (from Northern Gravel Company, Muscatine, IA) is shown in Figure 5.56 and Tables 5.8 and 5.9.

The slot size of the gravel packed well is selected so that it retains 90-95% of the gravel. The screen effective open area is determined same way, applying a

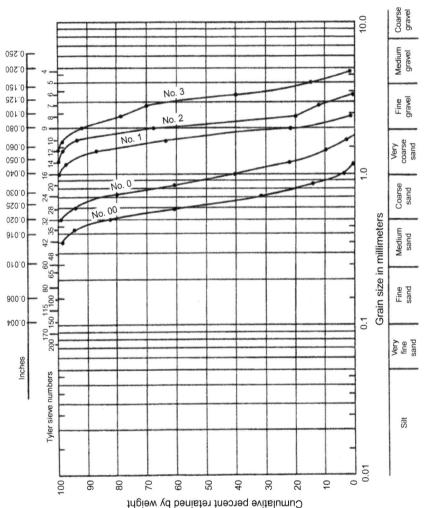

Figure 5.56 Grain size distribution curves for Northern Gravel Company well gravel packs.

Table 5.8 Approximate Size Ranges of Well Gravel Pack Material from Northern Gravel Company

No.	Size Range (in.)	Size Range (mm)
00	0.017-0.038	0.42-0.99
0	0.023-0.060	0.59-1.6
1	0.049-0.090	1.2-2.2
2	0.064-0.125	1.6-3.2
3	0.075-0.180	1.8-4.6

Table 5.9 Continuous Screen Slot Sizes Acceptable for Use with Well Gravel Pack Material from Northern Gravel Company, Muscatine, IA

No.	Slot Size
00	18
0	25
1	55
2	65
3	80

blockage factor of 50%. The screen entrance velocity is also determined based on the average of the formation and the gravel pack hydraulic conductivities. Most uniform gravel pack materials have higher conductivities (>6000 gpd/ft^2, which is at the top of the Table 5.6), which means that, for the smallest possible hydraulic conductivity (K) of the formation, the corresponding entrance velocity for a gravel packed well built in this formation will be 8 ft/min. Based on experience, if the formation has a K value less than 2000 gpd/ft^2, a safer and more realistic V_c can be obtained by doubling the formation's K value. The well screen length is based on the equation presented earlier.

The thickness of the gravel pack can vary. However, a too thick pack interferes with the development of the well and makes the well less efficient. Use of a pack that is too thick may result in the screen being exposed to the formation material, which causes sand pumping. Based on experience, the gravel envelope should be between 6 and 9 in. Thus, the borehole must be the casing diameter plus 12-18 in.

The filling depth of the gravel in the annulus (the space between the formation and the casing/screen) depends on the characteristics of the unconsolidated materials between the land surface and the well screen. The gravel pack is generally taken some distance above the screen to ensure the screen is not exposed to natural formation (especially of gravel settling during well development). Generally, 20-25 ft of gravel is needed above the screen.

If the gravel comes in contact with materials that are smaller than the diameter of the formation material for which the gravel pack was selected, fine particulates can migrate through the gravel to the well screen. In such cases, the gravel pack must be terminated before reaching the fine materials and sealed with bentonite or cement (see Figure 5.57).

After the pack design is complete, the results must be checked against the physical realities of tubular wells, as discussed previously. The approximate available drawdown should be determined and a maximum drawdown estimated to check if the design is reasonable and realistic under the existing

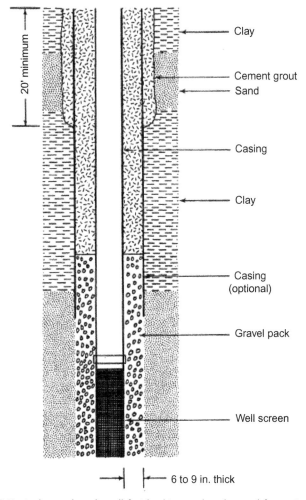

Figure 5.57 Typical gravel pack well finished in sand and gravel formations.

conditions. Often, conditions are encountered where both gravel–packed and tubular designs are possible. In that case, cost must be weighed to determine the final design.

The other type of well that is often used for water supplies is bedrock wells. These wells are constructed in fractured bedrock where the casing is generally installed 50 or 100 ft below the water table. An open hole, which contains the pump, is present below this level. However, such wells are not typically used in bank filtration systems.

Once a well is installed, development of the well must be done to remove the fine particulates and to enhance the specific capacity. This, in turn, increases the economic life of the well. Typical development procedures include pumping, surging, use of compressed air, hydraulic jetting, use of explosives, and hydraulic fracturing (for bedrock wells). Pumping and surging seem to be the most common techniques used in less developed areas. A well is pumped in a series of steps, and discharge is produced ranging from low discharge to a discharge greater than the design capacity. The intake of the pump is lowered to the middle of the screen zone to remove the most fine particulates. At each step, the water is pumped until it runs clear, and then the discharge rate is increased to the next higher level. This irregular and non-continuous pumping agitates the fine material around the screen. Coarser materials that fall into the well are removed by a bailer. Development by pumping is invariably the final procedure if other techniques are used.

Surging is another method used for developing water wells. A surge block is attached to a drill stem and moved up and down in a reciprocating motion (Figure 5.58). The gap between the well screen and the surge block is typically 2–5 cm. As the block is moved up and down, the water from the well is pushed into the formation via the gravel pack and then sucked back into the well bringing with it the fine particulates. Pushing the water into the formation may also break up any particle bridging. At the end, the collected fine particulates in the well are bailed out. Finally, a pumping procedure cleans up the fine particulates. Often, chemicals, particularly sodium hexametaphosphate (typically used to disperse clay), may be used in addition to surging.

After the well is developed and the pump is installed, a chlorination system is installed. Various chlorination systems are discussed in Sections 4.3 and 4.4. After the chlorination system is installed, the water quality should be tested and then supplied to the affected population.

If a large quantity of water is needed, multiple wells could be drilled on the bank of a river, and the pumped water from each well could be routed

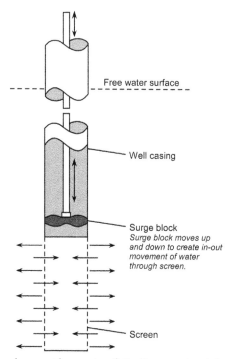

Figure 5.58 Well development by surging (http://www.ext.nodak.edu/extnews/snouts/ spout233.htm).

through one discharge pipe. In some riparian locations, where the depth of the surface to groundwater is relatively shallow (within the suction lift of a pump), a series of filter screens can be inserted and a centralized caisson can be used for pumping. For example, the Dusseldorf Water Works has a series of such filter screens, which have siphon tubes connected to a central pumping caisson (Figure 5.59).

Natural filtration has numerous advantages over the use of surface waters (Ray et al., 2002a). The pumped water from BF wells may not require any additional treatment besides disinfection. Even if a community rebuilds its water treatment plant after a disaster, conversion of a surface water intake system to a bank filtration system may reduce the costs associated with water treatment and remove a variety of contaminants that adversely affect water quality. As the particles are relatively low in count, need for coagulation and flocculation is reduced. Removal of natural organic matter also reduces the potential formation of trihalomethane in drinking water after chlorination. However, if the water contains significant amounts of dissolved iron or

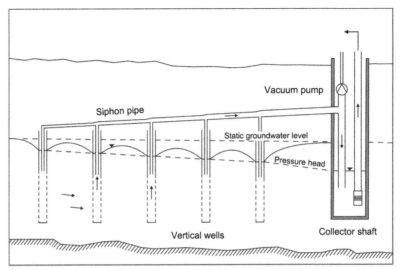

Figure 5.59 Siphon tubes connecting to pumpless wells (or filter screens); the water is pumped from the central caisson. *From Ray and Jain (2011).*

manganese, aeration may be needed to remove them. In addition, mineral removal depends on the hardness of the BF water.

Turbidity reduction through BF wells is sufficient to meet the drinking water standards in most locations without additional treatment. Figure 5.60 shows the turbidity of filtrate at Collector Well #1 of the Sonoma County Water Agency (SCWA) in Santa Rosa, California. The turbidity of the Russian River reached almost 500 NTU in the winter and early spring in both 2005 and 2006. During those periods, the filtrate from Collector Well #1 was still under 0.3 NTU. Figure 5.61 shows the river turbidity as well as the turbidity of filtration from the first well of the Louisville Water Company (LWC) in Louisville, Kentucky, in 2000. The river turbidity exceeded 800 NTU at some time during the monitoring period. However, the turbidity of the filtrate was around 0.2 NTU most of the time. Like the SCWA results, the LWC study showed that the spikes of turbidity in the river were not reflected in the filtrate from the wells.

The passage of river water through a BF well also helps stabilize the temperature of the filtrate. Figure 5.62 shows the variation in the temperature of the river water as well as the filtrate from the first collector well. This figure shows that the water temperature of the Ohio River varied between 2 and 33 °C while the filtrate temperature varied between 12 and 25 °C. There is a time lag between the high and low temperature values of the surface water

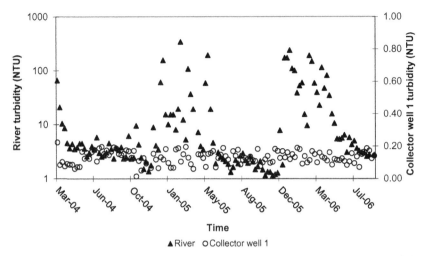

Figure 5.60 Turbidity of the Russian River and filtration from Collector Well 31 of the Sonoma County Water Agency at its bank filtration site monitored for a period of more than 2 years. *Redrawn using data presented in Ray and Jain (2011).*

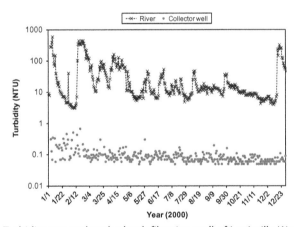

Figure 5.61 Turbidity removal at the bank filtration well of Louisville Water Company, Louisville, Kentucky, for the year 2000. *Redrawn using the original data used from Ray et al. (2002b).* (For the color version of this figure, the reader is referred to the online version of this chapter.)

and filtrate. These are shown by the dotted lines on Figure 5.62; which vary between 4 and 6 weeks. The viscosity of river water in winter months can be twice as high as it is in summer, which has several effects. First, well yield will decrease as the apparent hydraulic conductivity is reduced. However, in winter, the demand for water is usually low so this effect would not have any negative consequences. Second, if the surface water is used in a

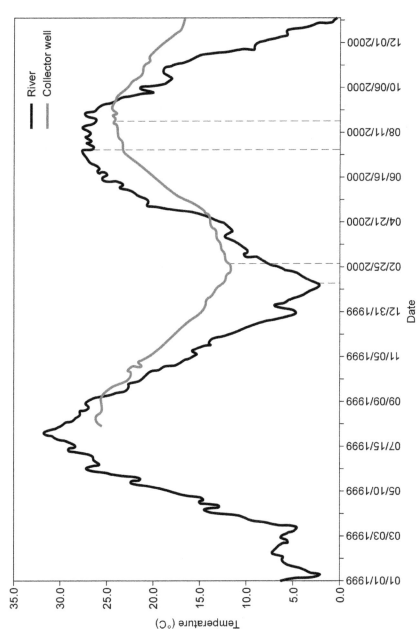

Figure 5.62 Temperature profile of water in the river and the collector well (bank filtrate) in Louisville, Kentucky. *From Wang (2002)*. (For the color version of this figure, the reader is referred to the online version of this chapter.)

treatment plant, the temperature variation of the water would require the operator to adjust coagulant dosing and other unit operations that are affected by temperature. The moderation of water temperature through bank filtration reduces these difficulties.

Predicting the temperature of water in bank filtrate is important to control processing and simulate chemical reactions because biogeochemical reactions are temperature sensitive. For example, degradation of dissolved organic carbon depends on temperature. The Dusseldorf Water Works in Dusseldorf, Germany, uses bank filtration to supply water to the city's residents. Figure 5.63a shows an aerial view of the river with the treatment plant on one side. Because the plant is on a bend, the bank closest to the treatment plant is steep, and it receives more sediments than the opposite bank, which causes clogging. Figure 5.63b shows a cross-section of the river near the treatment plant. Three monitoring wells are positioned near the production well. Well C is located 50 m away on the land side of the production well. Wells A and B are between the production well and the river. Each of these monitoring wells has the ability to sample water at different depths. (In the figure, the monitoring ports are marked 1 through 6, depending on the well.) The aquifer is layered at this site with hydraulic conductivity values ranging from 3.01×10^{-3} to 6.37×10^{-3} m/s from top to bottom. At the river-aquifer interface, there is a clogging layer with a hydraulic conductivity of 1.04×10^{-6} m/s near the bank and 9.26×10^{-4} m/s near the mid-section of the river.

Sharma et al. (2012) conducted a simulation of the temperature in the bank filtrate using heat as a reactive species in the multi-species transport model used. Figure 5.64 shows the temperature of the river water during a period of more than 1 year and the corresponding measured and simulated temperature values in port 1 of monitoring well B. As shown in this figure, the peak temperature in port 1 of this well was close to 10 °C cooler in summer, and the peak temperature of the bank filtrate was reached approximately 2 months after the peak temperature of the surface water was reached. The simulated and measured values are consistent.

Sharma et al. also conducted a simulation of the impact variable temperatures had on the concentration of oxygen in bank filtrate. In most isothermal models, the temperature is assumed to be a fixed number. Sharma et al. found that, by assuming a fixed temperature in the simulation domain (5, 15, or 25 °C), they were not able to match simulated oxygen concentrations to the observed concentration values at port 1 of monitoring well B (Figure 5.65). In this figure, the blue dotted line shows the temperature of the river water

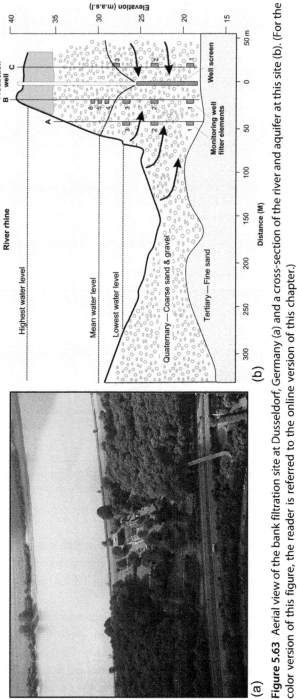

Figure 5.63 Aerial view of the bank filtration site at Dusseldorf, Germany (a) and a cross-section of the river and aquifer at this site (b). (For the color version of this figure, the reader is referred to the online version of this chapter.)

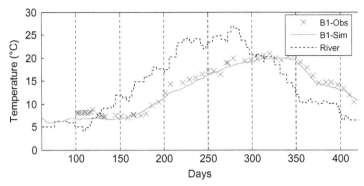

Figure 5.64 Observed temperature of the Rhine River and simulated temperature at port 1 of observation well B. *From Sharma et al. (2012)* (For the color version of this figure, the reader is referred to the online version of this chapter.)

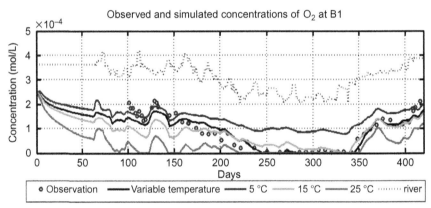

Figure 5.65 Simulated and observed concentrations of oxygen at port 1 of observation well B. *From Sharma et al. (2012).* (For the color version of this figure, the reader is referred to the online version of this chapter.)

over time. At an assumed 15 °C, the researchers' match was good between days 300 and 450; however, the fit was poor in the early part of the simulation. This figure also shows that for 100 days (between days 250 and 350), the bank filtrate nearly became anoxic.

In Figure 5.66, Sharma et al. show the difference in concentrations of oxygen, dissolved organic carbon, sulfate, and nitrate by assuming temperature-controlled degradation of dissolved organic carbon (DOC) and other reactions involving oxygen, nitrate, and sulfate. They also conducted simulations without considering the effect of DOC degradation

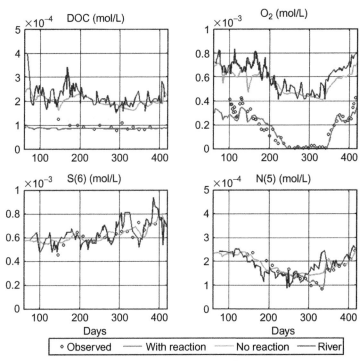

Figure 5.66 Simulated and observed concentrations of organic carbon, oxygen, sulfate, and nitrate for port 1 of observation well B. *From Sharma et al. (2012).* (For the color version of this figure, the reader is referred to the online version of this chapter.)

and temperature variations. The open circles are observed values, the red lines are values with reactions, the green lines are values without reactions, and the blue lines are the values in the river water. We find that if we do not include temperature-controlled reactions for DOC degradation, the predicted oxygen concentration in bank filtrate does not match observed values. However, as shown in Figure 5.64 as well as in the top right panel of Figure 5.65, if we do include temperature-controlled reactions, we see a good match between the observed and simulated values. In the lower right panel of Figure 5.65, we also see nitrate is substantially reduced during the period when oxygen is depleted. For sulfate (the lower left panel), a simulation with reactions was not any different from the one without reactions as the system did not reach a sulfate-reducing condition. In the top left panel, the measured and simulated DOC show that about 1 mg/L is nonbiodegradable. As the river water enters the aquifer, the biodegradable portion of DOC is consumed by bacteria through aerobic respiration.

A simultaneous decrease in oxygen occurs. As the oxygen is depleted between days 250 and 350, nitrate is used as an electron acceptor. As the nitrate is not fully removed, other redox reactions such as iron reduction or sulfate reduction do not occur. Thus, it is important that utility managers have an understanding of the effect of the redox conditions on resulting water quality.

Bank filtration systems also have other benefits. The formation of disinfection byproducts (DBPs) is an issue for water utilities that treat surface water with chlorine. The TOC in the river water is considered an indicator of the formation potential of DBPs. Bank filtration removes a portion of the TOC during the water's passage through the soil. At LWC, experiments were conducted at the first collector well to observe the amount of TOC removal when flow distance from the river–aquifer interface was increased. The collector well had seven laterals and four of them were directed toward the river (Figure 5.67). Lateral 4, which was directed toward the river and had a depth of 50 ft (15 m), was monitored, and piezometers were installed at depths of 0.6 m (W1), 1.5 m (W2), and 3 m (W3).

Figure 5.68 shows that the TOC concentration in the river varied from about 1.7–3.8 mg/L, with the peak occurring the late fall. Piezometers W1, W2, and W3 showed similar removal efficiencies, implying that a substantial amount of removal occurred in the first 0.6 m because of microbial reactions. The concentration of TOC in Lateral 4 was somewhat lower than that found at piezometer W3; this could be because of (a) additional degradation

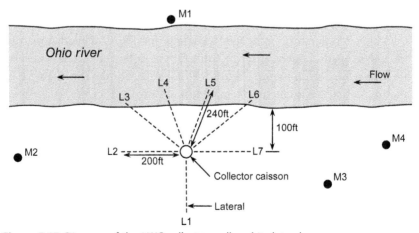

Figure 5.67 Diagram of the LWC collector well and its laterals.

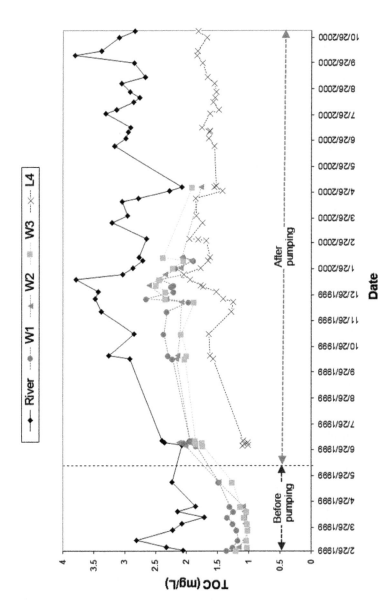

Figure 5.68 Removal of total organic carbon (TOC) from river water as the water moves from the river to Lateral 4 of the collector well. *Redrawn from Wang (2002).* (For the color version of this figure, the reader is referred to the online version of this chapter.)

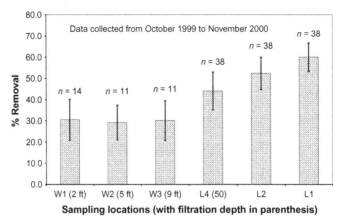

Figure 5.69 Reduction in TOC from the river water after passage through the aquifer. W1, W2, and W3 are piezometers and L1, L2, and L4 are laterals. *From Wang (2002).* (For the color version of this figure, the reader is referred to the online version of this chapter.)

of TOC during the water's passage through the aquifer, or (b) dilution with background groundwater. Typically, groundwater has a lower TOC than river water. Figure 5.69 shows the percentage of TOC reduction at the piezometers as well as Laterals 1, 2, and 4.

TOC removal at W1 (0.6 m) was about 30%, and this value did not significantly change until the depth of 3 m. However, TOC from Lateral 4, which was about 15 m below the riverbed, had a TOC that was 50% lower than that of the river water. As stated earlier, this could have been because of degradation of TOC or dilution with background groundwater. Lateral 1 extended away from the river and primarily pumped ground water. Its TOC was about one-third of that in the river. Lateral 2 received part groundwater and part river water.

Wang (2002) conducted formation potential tests for trihalomethanes, haloacetic acids (six of them), and total organic halogens on the water samples collected from the three piezometers (W1, W2, and W3) and Lateral 4 (Figure 5.70). The formation potentials were reduced between 25% and 50% in the waters from the piezometers and between 40% and 60% at Lateral 4. These values indicate TOC removal at these locations.

Figure 5.71 shows the log removal of total coliforms at LWC's collector well. The level of log removal varied from a low of 1 (90%) to a high of 5 (99.999%). Lower removals are typically associated with high flow events in rivers when the protective sediment at the river–aquifer interface gets

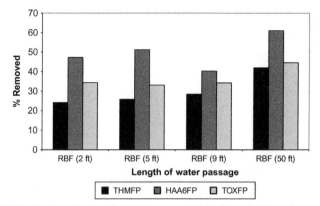

Figure 5.70 Reduction in the formation potentials for trihalomethanes (THMFP), six haloacetic acids (HAA6FP), and total organic halogens (TOXFP) at three piezometers and Lateral 4, which are at different depths from the riverbed. *From Wang (2002).* (For the color version of this figure, the reader is referred to the online version of this chapter.)

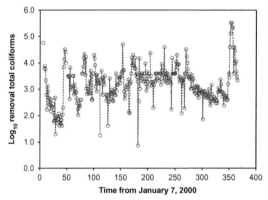

Figure 5.71 Log removal of total coliforms at the LWC collector well. *Redrawn using the original data from Wang (2002).* (For the color version of this figure, the reader is referred to the online version of this chapter.)

washed away. Figure 5.72 shows the removal of spores of *B. subtilis* at Laterals 1, 2, and 4 as well as the collector well. The river water had about 10,000 spores per 100 mL. Figure 5.73 shows the log removal values of these spores at L4 as well at the three piezometers (W1, W2, and W3). Between W1 and L4, the removals were between 1.5 and 3.5 log values.

Removal of microorganisms or other contaminants in river water can be low in collector wells compared to vertical wells in bank filtration systems. Typically, collector wells are built very close to a river, and they are pumped

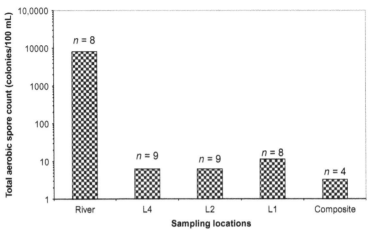

Figure 5.72 Removal of the spores of *B. subtilis* at individual laterals as well as at the collector well. *From Wang (2002).* (For the color version of this figure, the reader is referred to the online version of this chapter.)

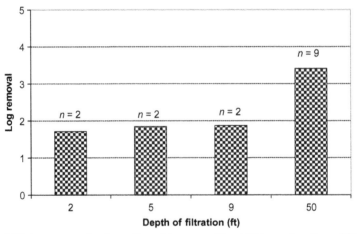

Figure 5.73 Log removal values of aerobic spores (*B. subtilis*) as a function of filtration distance. (For the color version of this figure, the reader is referred to the online version of this chapter.)

at rates much higher than vertical wells. As a result, the flow distance of the water is short. Removal rates for microbes can be much higher for vertical wells, which are typically pumped at lower rates and are built some distance away from the river. Therefore, the wells implemented in emergency scenarios should be vertical wells because they remove more pathogens and other chemicals.

Schijven et al. (1999) examined the removal of colliphages MS-2 and PRD1 in riverbank filtration wells (vertical wells) located in the lower Rhine Valley in the Netherlands. As the well distance increased from the river, so did the travel time of the water. These researchers observed an increase of more than 8 log removal values as the water's travel time increased to about 25 days (Figure 5.74). Medema et al. (2000) examined the removal of total coliforms and colliphages at a dune recharge site in the lower Rhine Valley in the Netherlands. They observed 4-6 log removals, depending on the organism (Figure 5.75).

Ray et al. (2002a) studied the removal of the pesticide atrazine at a collector well on the bank of the Illinois River that draws water for the city of Jacksonville, Illinois. Most of the time, the concentration of atrazine in the well was below detection, and the level in the river varied between 1 and 4 mg/L. In a flood event in the spring of 1996, the concentration of atrazine in the river water increased to about 11 mg/L (Figure 5.76). However, there was a small breakthrough in the well, and the concentration in the well increased to about 1.1 mg/L (about 90% removal). The current maximum contaminant limit (MCL) for atrazine is 3 mg/L. In essence, no additional treatment was required to remove the atrazine.

Schubert (2002) monitored levels of the complexing agent EDTA and the drug diclofenac at the bank filtration site in Dusseldorf. EDTA in the surface water varied between 6 and 12 µg/L (Figure 5.77) and that in the bank filtrate between 2 and 3 mg/L. The antibiotic diclofenac in the river

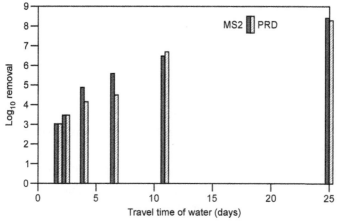

Figure 5.74 Removal of MS-2 and PRD1 phages from the Rhine River. *After Schijven et al. (1999).*

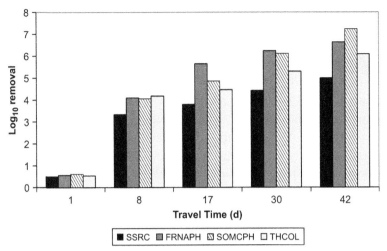

Figure 5.75 Removal of thermotolerant coliforms (THCOL), spores of sulfate-reducing bacteria (SSRC), somatic coliphages (SOMCPH), and F-specific RNA coliphages (FRNAPH) in the Netherlands. *After Medema et al. (2000).*

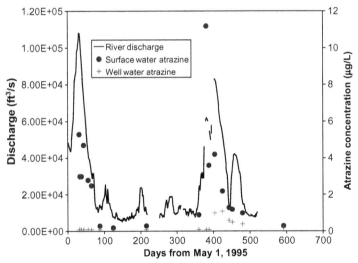

Figure 5.76 Concentration of atrazine in the Illinois River and in the collector well of the city of Jacksonville, IL.

water varied from 0 to 500 ng/L (Figure 5.78). Most of the time, the filtrate contained no diclofenac. However, in one incidence (during the peak event of the river), the concentration in the filtrate was about 40 ng/L.

It is clear that BF removes a variety of pollutants present in surface water. For proper operation of BF wells, it is essential that the river or stream has a

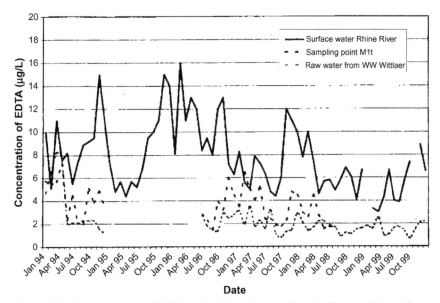

Figure 5.77 Concentration of EDTA in the Rhine River and in the filtrate collected from two locations. *After Schubert (2002).*

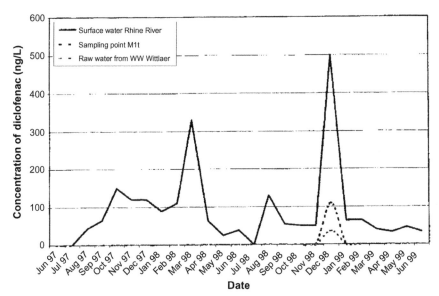

Figure 5.78 Concentration of diclofenac in the Rhine River and in the bank filtrate. *After Schubert (2002).*

high, but nonscouring, flow rate and clogging is minimal. Bank filtration systems located on the banks of stagnant rivers, such as those behind dams or diversion structures, seem to experience some water quality issues caused by deposition of organics (such as algae) on the riverbed. This leads to quick depletion of oxygen. In such cases, iron and manganese become soluble. Siting of the wells is also important from the water production perspective because a higher yield is possible when wells are located inside a river bend rather than on the outside (Figure 5.79). Figure 5.80 shows the flow path of water coming to a bank filtration well along the Elbe River. It seems most of the water comes from the other side of the river. In this particular case, the high nitrate level detected in the well was caused by water flowing from grape growing areas on the other side of the river even though the river nitrate level was lower than that of the pumping well.

In developing countries, the use of natural filtration has significant potential where the budget for building and maintaining water treatment

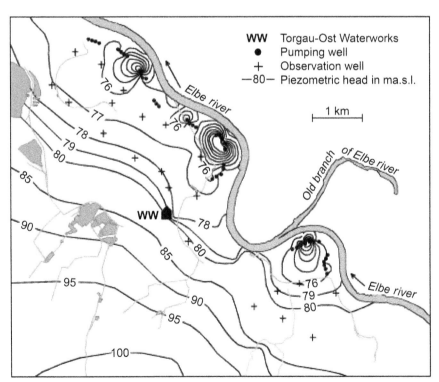

Figure 5.79 Location of wells inside bends along the Elbe River, Germany. *After Grischek et al. (2002).* (For the color version of this figure, the reader is referred to the online version of this chapter.)

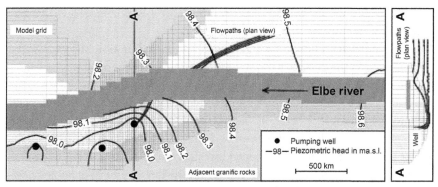

Figure 5.80 Flow paths of water pumped from a bank filtration well along the Elbe River near Dresden, Germany. Note that some of the water originates from the other side of the river. *After Grischek et al. (2002).* (For the color version of this figure, the reader is referred to the online version of this chapter.)

plants may not be able to keep up with population growth. Even if bank filtration is used only for pretreatment, the resulting water quality will be far superior to that of the surface water. Natural filtration can also defer expansion costs of water treatment plants. Appropriate technologies can be adapted to local situations to make bank filtration systems work. For example, Koop wells used in the hilly state of Uttarakhand where filter and a piece of geotextile are used to wrap the screened pipes, to produce filtrate (Sandhu et al., 2011). This also increases the travel time of the water from about 5 min to more than 4 h. In field settings, the placement of wells and pumps should be optimized so that the travel time is not too short to reduce the contaminants in the river water.

Developing countries such as China, Egypt, and India face tremendous challenges in keeping their rivers clean because of industrial discharge. In such situations (where the source water is poor quality), the bank filtrate probably cannot be supplied to consumers without additional treatment. Many water treatment plants in these countries pretreat water with chlorine to get rid of color and then disinfect to maintain chlorine residuals in the system. Rarely reported or recognized, however, is the positive effect BF has on minimizing the formation of carcinogenic disinfection byproducts.

Ray (2008) summarized the worldwide potential of riverbank filtration: Vast areas of the world (Asia, Africa, and South America) are still untapped territories in which to build bank filtration systems. However, Sandhu et al. (2011) highlight challenges that a bank filtration system can face in a developing country such as India. Bank filtration is rapidly expanding in South

Korea, where the technology has been found to be an inexpensive alternative to the membrane treatment process.

The potential climate change effects on bank filtration should also be examined. As the climate becomes more erratic (warmer summers or colder winters) and low-flow periods in river systems become prolonged, the impact of waste discharge to rivers can adversely affect the operation of bank filtration systems. Schoenheinz and Grischek (2010) have pointed out the potential climate change effects on bank filtration. Eckert et al. (2008) highlight the impact of climate-induced low flows in the Rhine River as well as the impact of warm water temperatures on the development of anoxic conditions at Dusseldorf's bank filtration facility.

Floods generated by heavy rainfalls can cause a lot of diffuse pollutants to enter rivers in agricultural watersheds, especially in spring when snow is also melting. Bacteria, pesticides, synthetic and dissolved organics, and nutrients are the primary contaminants during this time. In flood situations, finding supplies of electric power as well as preventing pump motors from being submerged can be issues for bank filtration systems. Thus, it is advisable to build a well house on high ground and put berms of embankments around a chlorine contactor.

CHAPTER 6

Emerging Technologies for Emergency Applications

Contents

6.1 Nanotechnology   169
6.2 Renewable Energy   170
6.3 Iodinated Resins   172

> **Keywords:** Emergency, Nanotechnology, Carbon nanotube, Renewable energy, Iodine

Emerging water treatment technologies currently being researched and developed will play an important role in the future of emergency water treatment devices. Several of these technologies are being developed, such as carbon nanotubes, or are already in use and are being refined, such as iodinated resins and renewable energy sources. All of these emerging technologies offer a form of improved water treatment, whether that improvement is a new energy source or fewer disinfection byproducts.

6.1 NANOTECHNOLOGY

Nanotechnology has great potential in water treatment applications. Nanomaterials are superior filters and adsorbents because they have extremely small pore sizes (1-100 nm) and large surface areas. Also, some nanomaterials exhibit oxidative/reductive properties that can be used in the degradation of chemical pollutants. Nanomaterials that have high antibacterial properties include chitosan, metallic nanoparticles (e.g., silver ion), photocatalytic TiO_2, nanofilters (NF), aqueous fullerene nanoparticles, and carbon nanotubes (Ngwenya et al., 2013). These superior physical, chemical, and antibacterial properties, combined into one process, have the potential to provide low cost and effective full-scale water treatment (Brame et al., 2011) as well as water treatment for emergency applications. Nanotechnology water treatment could be implemented in developing countries where infrastructure is limited and water is scarce. While the benefits of using

---

☆"To view the full reference list for the book, click here"

nanoparticles may seem enticing, negative environmental effects are being investigated. For example, using metal-based nanoparticles could create antibiotic-resistant bacteria strains (Brame et al., 2011). Few studies about the environmental effects of using nanomaterials exist; thus, more studies are needed before nanomaterials can be widely implemented.

Carbon nanotubes are an emerging form of nanosorbent that can be used for emergency water treatment applications. Carbon atoms can be stacked in a hexagonal pattern and then rolled up into a tube to form a nanotube. These tubes have great potential for use in water purification because they can remove contaminants such as bacteria and viruses through adsorption and size exclusion as carbon nanotubes are $10,000 \times$ thinner than a human hair (U.S. Army Public Health Command, 2010). Carbon nanotubes are ideal for emergency situations because they have a very low propensity for fouling so they work efficiently in a wide variety of source waters and require little maintenance. Very little energy is required to pressurize flow through the nanotubes, enabling the use of gravity-driven flow (Brady-Estevez et al., 2010; Rahaman et al., 2011). Carbon nanotubes can be incorporated into membranes as a nanomesh or used in conjunction with membranes to enhance water treatment. An example of a technology that uses a carbon nanomesh filter is the Seldon WaterBox, as discussed in Section 5.3.3 (Jones, 2011; Seldon, 2012). To increase the microbiological effectiveness of a carbon nanotube (CNT), silver particles can be impregnated into the CNT to increase its disinfecting capabilities (Lukhele et al., 2010). As research and development of nanotechnologies increase, the use of carbon nanotubes will probably also increase. However, current production techniques require more development before carbon nanotubes can be used for emergency water treatment (U.S. Army Public Health Command, 2010).

6.2 RENEWABLE ENERGY

Renewable energy is a highly useful power source during an emergency. During chaotic times when traditional power sources such as electricity and diesel fuel may be unavailable, energy from the sun and wind can be harnessed. Solar and wind energies are two of the main renewable energies that are used to provide drinking water. These renewable energies can be used in emergencies to provide water by pumping clean groundwater or through various other water purification processes. While renewable energy benefits the environment because it is a clean energy, and it can be used when traditional forms of power cannot, renewable energy devices require a large initial investment for the photovoltaic (PV) cells and connecting technology. Currently, the only way renewable energy can be cost effective is if it's used in a region that has a large solar capacity and high costs for diesel fuel (Abraham and Luthra, 2011; Bilton et al., 2011).

Many case studies have examined the use of renewable energies to provide drinking water in developing countries, and a growing number of studies (currently very few) analyze the use of renewable energy technologies during the initial, acute phase of emergencies. Because renewable technology devices may be larger, more difficult to transport, and require a specific set up as well as long-term maintenance, they may be more useful for a long-term, sustainable response to an emergency. Many photovoltaic cells and wind turbines have a lifespan of 20-25 years and can provide water for a community long after an emergency is over. Both pumping systems and treatment systems can be scaled up for larger capacity or scaled down for a smaller capacity, depending on the application. Providing energy for a larger system merely requires the addition of more PV cells. Renewable energy has been employed in both rural, off-grid applications for small communities as well as in large-scale water treatment systems. Power requirements for water treatment include pumping, UV disinfection, membrane filtration, and desalination/reverse osmosis (RO) units.

Various scales of photovoltaic-powered membrane filtration water purification systems exist. RO is a popular technology for areas with brackish water and has been examined for use in both community-scale and relatively large plants. Membrane desalination is another form of purification that can be used to treat brackish water. The main disadvantage of using RO or other desalination processes is the amount of energy that's required to run these systems. PV cells can provide enough energy for desalination if the treatment unit is small, but as the capacity of the plant grows, the process becomes more expensive. The more energy needed for the plant, the larger the area needed for PV cells. Small, personal-size RO units powered by manual pumps or PV cells can be stored on life rafts for emergency situations (Katadyn, 2013).

Wind energy is another popularly used form of renewable energy. Wind energy technology is fairly well developed and relatively cheap (Miranda and Infield, 2002). Like solar technology, wind energy is most cost effective when used in off-grid applications where traditional forms of energy are not available. Many studies have examined using wind technology for applications such as pumping water for rural irrigation (Heijman et al., 2009; Salomonsson and Thoresson, 2010) and treating brackish water or seawater to make potable water (Gauto, 2012; Liu, 2009; Park et al., 2010). The difficulty with using wind as a power source is that wind is random and variable. RO systems require a certain amount of power that may not always be achieved because of the randomness of the wind. Park et al. (2010) found that RO systems could treat low-salinity source water using wind energy at high wind speeds, but intermittent operation still needs to be studied. The lack of consistency in wind power requires either short- or long-term storage of either electricity or water to provide a constant flow of drinkable water. Batteries for electrical storage can greatly increase the initial investment costs, so research has been conducted to find power-saving alternatives, such as energy-recovery pumps (Heijman et al., 2009).

Occasionally, wind and solar energies have been combined to provide extra energy and redundancy for water treatment systems. Vick and Neall (2012) found that using hybrid wind and solar power provided enough energy to pump water during low water times, but the combination provided too much energy during wet seasons, resulting in excess pumping. These authors suggested that this excess energy could be used for other applications, such as heating a tank of water or, if the water needs to be purified, UV disinfection. Vitello et al. (2011) tested a mobile UV disinfection system that used a combination of solar and wind power supply. Like Vick and Neall (2012), these researchers found that the wind turbine provided a redundancy that was unnecessary for most situations, and removing the wind turbine drastically reduced the cost of the system. Both Vitello et al. (2011) and Vick and Neall (2012) tested off-grid systems that had capacities ranging from 9 to 40 L/min, enough for a small community water supply. Another example of hybrid solar and wind energies is the Solar Cube, a product by Spectra Watermakers (2013), which combines solar and wind energies to power a large, portable RO unit that uses a special pump to reduce the draw of power from the renewable energy source. This unit has a capacity of 3500 L/h, but it requires a relatively large investment.

6.3 IODINATED RESINS

Iodine has been used for disinfection of water since World War I. Incorrect doses of iodine result in high toxicity and irritation; thus, iodine is used less frequently than chlorine to treat water. However, iodine does not produce a bad taste as chlorine does, and iodine is an effective microbial disinfectant, so its use continues to be explored (Mazumdar et al., 2010; Ngwenya et al., 2013). Different ways to "tame" iodine have been investigated in various studies (Mazumdar et al., 2010). NASA uses an iodine resin in their water treatment system to reduce microorganisms (NASA Spinoff, 1995). A product called the Survival Bag employs this process, using a filter with iodine resin to inactivate microorganisms and then an iodine scavenger to remove any iodine species and reduce toxicity related to iodine (World Wide Water, 2013). Iodine disinfection continues to be researched as it has the potential to produce harmful byproducts that are similar to chlorination byproducts. Research has found that, while an iodine solution produces a large amount of byproducts, products that used resins were not as likely to produce harmful byproducts (Mazumdar et al., 2010).

CHAPTER 7

Water Infrastructure Development for Resilience

Contents

7.1 Need for Water Infrastructure Development 175
7.2 Infrastructure Improvements for Developed Countries 176
7.3 Infrastructure Improvements for Developing Countries 177
    7.3.1 Urban Areas: Conventional Water Treatment Plants 178
    7.3.2 Peri-Urban and Rural Areas: Small-Scale Treatment Systems 178
    7.3.3 Rural Areas: Point-of-Use Treatment 179
7.4 Short-Term Solutions 179
7.5 A Wholesome Approach to Infrastructure Development 180
    7.5.1 Government Involvement in Infrastructure Development 181

Keywords: Water infrastructure, Infrastructure development, Point of use, Developing Countries, Developed countries

7.1 NEED FOR WATER INFRASTRUCTURE DEVELOPMENT

With a consistently increasing demand for water (Foster et al., 2012) and a rise in the number and severity of extreme atmospheric events (Hill et al., 2012), the likelihood that water supply infrastructures will be damaged by extreme weather-related events is increasing. It is very important that countries develop their water infrastructures accordingly. While wealthy countries are able to make significant investments in disaster preparedness and to increase the structural resilience of their water systems to reduce the impact of weather-related events (Hill et al., 2012), developing and poor countries lack the infrastructure to manage, store, and deliver water resources (Grey and Sadoff, 2007) and lack the funding to invest in this infrastructure. Developing countries often have extreme weather and sometimes "difficult" hydrology, a condition characterized by susceptibility to severe flooding or drought. Poor countries make the lowest infrastructure investments and have the weakest institutions (Grey and Sadoff, 2007); thus, there are more risks to their water systems. Because of this lack of infrastructure, poor countries are the most vulnerable to disasters as they are unable to adjust

☆"To view the full reference list for the book, click here"

for hydrological variability (Grey and Sadoff, 2007) and do not have the money to deal with the hidden costs of infrastructure damage (Hill et al., 2012). Unable to bear the costs of repairing and maintaining infrastructure, developing countries face barriers to both short-term recovery and long-term development of their water systems (Hill et al., 2012).

7.2 INFRASTRUCTURE IMPROVEMENTS FOR DEVELOPED COUNTRIES

Developed countries have the resources to be responsive and resilient during times of emergency. With developed water infrastructures and individuals dedicated to maintaining and operating water utilities, these countries can adequately prepare for emergencies. Emergency plans in developed countries usually include preparation; plans for emergency, short-term response; and plans for long-term, sustainable response. Emergency preparedness includes testing scenarios based on disaster-prone areas and water supply system vulnerabilities (Patterson and Adams, 2011). Emergency preparedness actions, including infrastructure strengthening and utility training (Patterson and Adams, 2011) and pre-positioning of supplies (Crowther, 2010), can be based on these scenarios. Water infrastructure strengthening could include:

- Reinforce well houses and pump stations
- Sandbag critical water and wastewater components, such as pumps and building entrances
- Overchlorinate water supplies to protect against waterborne pathogens
- Top off water storage tanks and close main valves in anticipation of pipe breaks
- Set electric components to manual mode
- Isolate or shutdown exposed pipes at river crossings (adapted from Patterson and Adams, 2011)

Training sessions for utility operators and managers can include a focus on specific scenarios to facilitate responses to different types of emergencies and should include instruction about proper maintenance of the system (Patterson and Adams, 2011). Another emergency preparedness action for developed countries includes pre-positioning of emergency drinking water supplies, which can guarantee potable water for affected areas. However, the most effective locations for infrastructure strengthening, pre-positioning of supplies, and emergency procedures are based on the outputs of the emergency models and scenarios. Inaccurate emergency prediction models result in inefficient use of pre-positioned supplies, which is not cost effective.

Improving modeling of disasters and emergency predictions will increase the effectiveness of emergency preparations, as even a small change in a forecast can result in a large change in a disaster's area of impact (Crowther, 2010).

In the short-term, immediate response to a disaster, it is important to assess the public water system and prioritize facilities for repair and technical assistance. Crews that perform these technical assessments and analyses can be formed before an emergency and evaluate water treatment facilities during the disaster or immediately afterward (Patterson and Adams, 2011). During the initial phase of the emergency, resources should be allocated as quickly as possible (Oloruntoba, 2010). After the acute emergency phase has passed, a long-term response to the emergency is needed. The long-term emergency response should include an evaluation of system integrity as well as the potential for long-term improvements.

While developed countries may have more resources to provide aid during emergencies than developing countries, developed countries can still improve their response times, management of water supply systems, and communication techniques, especially during the initial phases of disasters (Oloruntoba, 2010; Seyedin and Jamali, 2011). For example, during Hurricane Katrina, the emergency response was robust but was not enough to meet the emergency demands (Tsai and Chi, 2012). In contrast, Australia applied a vast amount of resources during the very early stages of Cyclone Larry, resulting in a more comprehensive emergency plan (Oloruntoba, 2010); thus, developed countries need to improve their emergency responses.

7.3 INFRASTRUCTURE IMPROVEMENTS FOR DEVELOPING COUNTRIES

Emergency response phases in a developing country are similar to the phases in developed countries, though less sophisticated. Developing countries lack infrastructure and often mismanage resources, which worsens the consequences of disasters. During disasters in developing countries, water from any source is used to sustain life, whether or not that source is safe. Relief in the form of water supplies or treatment can be provided by Nongovernmental organizations (NGOs) and local or nonlocal governments. Relief organizations choose the types of water treatment devices they feel are most appropriate for each emergency and location. While creating and strengthening infrastructure may be an emergency preparedness goal for developing countries in the future, doing so immediately may not be cost effective (Crowther, 2010; Lougheed, 2006) as constructing pipelines and treatment infrastructure requires a huge investment and a continual input of resources (Reiff et al., 1996), and building infrastructure does not guarantee that people will consume clean water (Arvai and Post, 2012). Additionally, homes in rural areas are usually widespread, making infrastructure development impractical. Even a centrally located improved water source would require a widespread population to travel great distances to access the water, decreasing the effectiveness of the infrastructure development. Instead, improving the existing, different types of treatment systems could be an intermediate goal (Reiff et al., 1996; Swartz, 2009).

7.3.1 Urban Areas: Conventional Water Treatment Plants

The United Nations has predicted that 56% of people in developing counties will live in urban areas by 2030 (Lee and Schwab, 2005). As populations in urban areas continue to grow, the amount of water treatment needed in urban areas will also increase. Urban areas typically have water infrastructures consisting of conventional treatment plants and the pipelines that convey the treated water to the community. While a water treatment system in conjunction with a piping infrastructure could provide large quantities of water to urban residents, this tap water might not be good quality because of inadequate water treatment or failures in the distribution system (Lee and Schwab, 2005; Rosa and Clasen, 2010). This lack of quality could be caused by bacterial growth, which, in turn, could be caused by interrupted service (where customers receive water only certain hours of the day), negative hydraulic pressures in pipes, infrastructure aging, and improper disinfection techniques (Lee and Schwab, 2005). Improvements that could be made to reduce the contamination within an urban water system include chlorination at multiple points within the distribution system, leak detection and prompt repair, rehabilitation of old pipes, and routine checks on valves and other preventative maintenance (Lee and Schwab, 2005). Making these improvements to urban water systems may not prevent disasters from contaminating the water in the system's infrastructure but would make these systems more resilient to such damage.

7.3.2 Peri-Urban and Rural Areas: Small-Scale Treatment Systems

In a peri-urban area, the source of water could be a small treatment plant that is supplied with surface water or groundwater. Makungo et al. (2011) noted that 20% of the population in South Africa is serviced by small water treatment plants. Populations that are served by small-scale water treatment technologies may have difficulty obtaining an adequate quantity and quality of drinking water even when there is no emergency because they lack experienced water managers and efficient systems (Makungo et al., 2011). To improve these small-scale water treatment plants, training should be provided to ensure that operators use the correct chemical doses and monitor effluent quality. After the treatment technologies have been upgraded sufficiently, utility managers should be trained in how to keep the plant functioning during an emergency situation.

7.3.3 Rural Areas: Point-of-Use Treatment

Point-of-use (POU) water treatment is a commonly employed form of water treatment in rural areas of developing countries where piped and other improved water supplies are unable to reach residents (Rosa and Clasen, 2010). Although it is difficult to measure the efficiency and efficacy of POU water treatment devices because of differences in metrics, POU has been shown to be one of the most cost effective ways to treat water supplies in developing countries (Rosa and Clasen, 2010). Of all the water treatment technologies used in developing countries during times of nonemergency, POU treatment is probably the most applicable treatment technology during times of emergency. Urban water infrastructures can fail during emergencies as treatment facilities can become overwhelmed, managers and operators of water treatment plants can be injured, and groundwater wells can become contaminated. Treatment facility vulnerabilities can leave citizens with no safe drinking water alternatives. However, POU technologies allow the affected population to treat their own water. Therefore, POU technologies should become widespread not only in rural areas but also in urban areas. If use of POU technology was increased, supply chains for POU devices would already be well established before an emergency (making it easier for communities to obtain supplies during emergencies), and locals would already have a working knowledge of how POU devices operate.

7.4 SHORT-TERM SOLUTIONS

POU and packaged technologies (see Chapter 5) are not only applicable in developing countries, but they can also be used in developed countries during emergencies. Because POU technologies allow an individual or small community to treat their own water, these technologies free people from dependence on conventional treatment systems, which are susceptible to failure at multiple points. POU technologies can be integrated into emergency preparedness plans in developed countries through pre-positioning. Instead of pre-positioning supplies such as bottled water, POU treatment devices, which can be used for the short-term phase of an emergency (until conventional treatment is operational again), can be pre-positioned. Crowther (2010) discusses an incentive or reimbursement program that the government can offer to encourage pre-positioning of POU technologies. This prepositioning may be difficult as it involves predicting the location and intensity of the disaster (Crowther, 2010).

7.5 A WHOLESOME APPROACH TO INFRASTRUCTURE DEVELOPMENT

While the above recommendations for rural and urban areas are short-term solutions based on current conditions and needed improvements, making these changes is easier said than done. These changes have not already been implemented because of politics, social factors, and economics. Lee and Schwab (2005) noted that water infrastructure collapse has been associated with lack of political support, inadequate or improper use of funds, poor management, and poor cost recovery. Other contributing and underlying factors include poor communication, insufficient community involvement, inadequate human resources, or lack of trained personnel. All these problems must be addressed to create an effective and sustainable water supply distribution system as correct maintenance and operation contributes to a water treatment system's lifespan.

To develop a sustainable approach to water supply, there needs to be an inclusive watershed management policy that will eventually decrease the cost, improve the quality, and increase the quantity of drinking water. Through watershed protection, water sources will become less polluted and easier to treat. Proper sanitation services and hygiene will contribute to reducing pollution. Proper maintenance of the water distribution system will promote the longevity of the infrastructure. Minimizing water waste (known as "unaccounted-for water") is also an important step in increasing water availability (Lee and Schwab, 2005). As unaccounted-for water use in developing countries is high because of leaks or illegal connections (an average of 30-50% of the water treated in developing countries), reducing this water waste would result in a larger amount of water being available to those who may not have had access to water previously (Lee and Schwab, 2005). Increasing the accessibility of water would result in large financial benefits and increase the standard of living in the watershed. Although a large financial gain is expected from increasing access to drinking water, an initial investment is needed to improve the water system infrastructure. While funding this initial investment may be a challenge for developing countries, the expected financial gain may help the country decide to improve its water system infrastructure.

7.5.1 Government Involvement in Infrastructure Development

According to Makwara (2011), water is both a social and economic good, a basic right to which every human deserves access. However, water cannot be treated as purely an economic good or purely a social good but must be treated as some combination of the two. Infrastructure development and investment must reflect this concept, making water more affordable and readily accessible to all individuals after emergencies. The government is primarily responsible for developing resilient water policies. Government must represent all citizens of the country and protect and guard the country's water resources. To develop these resilient water policies, an adequate cross-sector dialogue must be created within the government to fully integrate water resource management into national policy (Foster et al., 2012). Increasing water infrastructure resilience will require careful planning and use of available hydrogeological information. In general, government involvement in improving water resource infrastructure should include policies such as: water rights, permits or allocations, the ability to impose bans, charging fees to cover monitoring costs, and sanctions for noncompliance (Foster et al., 2012). Water policy and the direction of infrastructure developments should contribute to national goals and include inherent value beyond the project's main purpose. For example, watershed management policy could include the development of roads, dam construction, reforestation, revegetation, erosion control, and irrigation system improvement (Lambert et al., 2012). All these projects would add to the resilience of the water infrastructure and contribute to the development of the country's economy.

Tables 7.1 and 7.2 are adapted from Peter-varbanets, 2009. Table 7.1 summarizes the types of emergency water treatment that are currently being used or being developed. This table enables the reader to understand an overview of all the different types of water treatment techniques that are discussed in this book. Table 7.2 is a more detailed look at emergency water treatment and only includes packaged systems. These tables assist in the selection appropriate technologies for emergency water supply. (Peter-varbanets, 2009).

Performance: "++": the water produced is microbiologically safe according to WHO standards if the treatment is performed correctly; "+": the water produced is safe only under certain conditions (e.g., if raw water is not turbid), or the system is efficient against most microorganisms with few exceptions.

Ease of use: "++": daily operation is limited to hauling raw water and collecting treated water; "+": requires additional (time consuming) operations that may be performed by unskilled person with little or no training.

Sustainability: "+": the system may be produced locally from locally available materials, with limited use of chemicals and nonrenewable energy sources; "−": the system requires chemicals or nonrenewable energy sources for daily operation; "−−": widespread application causes or may cause in future significant environmental damage (e.g., deforestation caused by cutting trees to make fuel to boil water).

Social acceptability: "++": the application is based on tradition or it is already in use; "+": available studies showed adequate levels of social acceptance; "+/−": available studies are contradictory, or the results depend on the region studied.

The investment costs listed in Table 7.1 generally include the costs needed to buy, deliver, and install the system. The operational costs include the costs of reagents, energy, and servicing if needed, as well as maintenance and replacement parts.

POU = Point of Use

SSS = Small scale system, serves a small community

DC = Developing Countries

Table 7.1 Overview of Types of Emergency Water Treatment

Technology	Type of Supply	Investment Cost	Maintenance Cost	Performance	Ease of Use	Maintenance	Sustainability	Power Needed	Social Acceptability	Source
Boiling	POU	Cook pot and fuel price	Depends on fuel price	++	+	Fuel collection	– –	Fuel	++	Peter-Varbanets (2009)
Bottled Water	For a family of 4 per year	Depends	Depends	Depends	Depends on delivery distance	Replenish supply	–	None	+	Peter-Varbanets (2009)
Sopas/Sodis	POU	Plastic/glass bottles	None	+, low turbidity	+, training needed	Clean regular, time consuming	+	Solar	–/+	Peter-Varbanets (2009)
Biosand Filters	POU	$10–20	None	+/– (minimal virus removal)	++	Clean every few months	+	Gravity	+	Peter-Varbanets (2009)
Coagulation/filtration	POU	$5–10	$140–220	+	+, training needed	Clean regular, time consuming	–	None	–/Taste	Peter-Varbanets (2009)
Free Chlorine	POU	$2–8	$1–3	+	+	Regular	–	None	–/Taste	Peter-Varbanets (2009)
Ceramic Filters	POU	$8–$200	$2–$12	+/– (minimal virus removal)	++	Cleaning and replacement	+	Gravity	+	Peter-Varbanets (2009)

Continued

Table 7.1 Overview of Types of Emergency Water Treatment—cont'd

Technology	Type of Supply	Investment Cost	Maintenance Cost	Performance	Ease of Use	Maintenance	Sustainability	Power Needed	Social Acceptability	Source
Forward Osmosis	POU	$2-300	$2.30/indiv. refill	++	+	Cleaning and refilling	–	Osmotic pressures	?	HTI (2010)
Micro-filtration	POU and SSS	$6-$120	?	+/– (minimal virus removal)	+/++	Cleaning and replacement	–/+	Gravity	–/+	Lifestraw (2008) and Peter-Varbanets (2009)
Ultra-filtration	POU	$20-$26,000	NA	++	++	Backflushing	+	Gravity, pumps, solar	+	Lifestraw (2008), Sunspring, and Prefector -E
Reverse Osmosis	POU	$220	?	++	++	?	–/+	Manual or Car Battery	+/–	Ray et al., (2012)

Adapted from Peter-Varbanets et al. (2009).

Table 7.2 Available Packaged Filtration Systems

Membrane Technology	Type of Supply	Capacity (L/d)	Pre or Post Treatment Needed	Feed Water Quality	System Investment	Maintenance/ Operation	Energy Required	Application
RO	Packaged and portable	136–170	None	Brackish to fresh water	~$220	Replace filters, clean hoses	Manual bike pump or car battery	Tested with military
UF, POU	Lifestraw Family	20–30 or 18,000 L/ system	Chlorine	Surface or groundwater	$40	Daily backflushing	Gravity	Tested in DC
UF, POU	Lifesaver Jerrycan	2 L/min or 15,000 L/ system	None	Surface water	$270–332	Clean regularly	Manual hand pump	Applied in DC
UF, SSS	Sunspring	7000–19,000	None	Surface or well water	$25,000	Minimal	Solar panels	Applied in emergencies
UF, SSS	"Arnal" System	1000	Coarse filter, microfilter, security filter	Surface water	NA	NA	Manual rotation wheel or generator	Tested in DC
UF, SSS	Perfector E (Norit)	48,000	Multistage, MF, UV	Brackish, highly polluted water	$26,000	Maintenance on a long term	Fuel	Applied during emergencies
UF, SSS	WaterBox	2 L/min or 30,000 L/ system	None	Surface water	$3190–$7975	Replace filters when needed	AC/DC Power	Military use, Applied in emergency in DC

Continued

Table 7.2 Available Packaged Filtration Systems—cont'd

Membrane Technology	Type of Supply	Capacity (L/d)	Pre or Post Treatment Needed	Feed Water Quality	System Investment	Maintenance/ Operation	Energy Required	Application
UF, SSS	SkyHydrant	5–00 L/h	Chlorine if being stored	Surface water, no salt removal	>$1000–2000	Backwash daily, chemical clean monthly	Gravity, 4 m of head	Applied in DC and emergencies
UF, SSS	iWater Cycle	400–900 L/h	None	Surface water	~$3000c	Backwash	Human riding a bicycle	Applied in Emergency, DC
MF, POU	FilterPen	100 L/ system	None	Surface water	$50	Disposable	Human powered (suction)	Applied in DC
MF, POU	Lifestraw Personal	700 L/ system	None	Surface water	$3–$20	Disposable	Human powered (suction)	Applied in DC
MF, POU	Katadyn	100–750 L/ system	MF, active	Surface water	$200–400	None, cleaning	Manual pump	Applied in DC
MF, POU	Ceramic Candles	10,000 L/ system	None	Surface water	$4–25	None or cleaning	Gravity	Applied in DC

Adapted from Peter-Varbenets (2009).

REFERENCES

Abbaszadegan, M., Hasan, M.N., Gerba, C.P., Roessler, P.F., Wilson, B.R., Kuennen, R., Van Dellen, E., 1997. The disinfection efficacy of a point-of-use water treatment system against bacterial, viral and protozoan waterborne pathogens. Water Res. 31, 574–582.

Abraham, T., Luthra, A., 2011. Socio-economic & technical assessment of photovoltaic powered membrane desalination processes for India. Desalination. 268 (1–3), 238–248.

Alcolea, A., Renard, P., Mariethoz, G., Bertone, F., 2009. Reducing the impact of a desalination plant using stochastic modeling and optimization techniques. J. Hydrol. 365, 275–288.

Al-Hayek, I., Badran, O.O., 2004. The effect of using different design of solar stills on water distillation. Desalination. 169, 121–127.

Almas, L.K., 2006. Cost comparison of water treatment systems to improve water quality for municipal use. 2006 paper 45, http://opensiuc.lib.siu.edu/cgi/viewcontent.cgi?article=1063&context=ucowrconfs_2006.

Amazon.com, 2014. Lifestraw 1.0 Water Purifier. http://www.amazon.com/LifeStraw-Family-1-0-Water-Purifier/dp/B00FM9OBQS.

Andreatta, D., 2009. The solar puddle. Solar Cookers International (SCI). solarcooking.org/pasteurization/puddle.htm. (accessed July 5, 2009).

Arnal, J.M., Fernindez, M.S., Verdti, G.M., Garcia, J.L., 2001. Design of a membrane facility for water potabilization and its application to Third World countries. Desalination. 137, 63–69.

Arnal, J.M., Garcia-Fayos, B., Verdu, G., Lora, J., 2009. Ultrafiltration as an alternative membrane technology to obtain safe drinking water from surface water: 10 years of experience on the scope of the AQUAPOT project. Desalination. 248, 34–41.

Arvai, J., Post, K., 2012. Risk management in a developing country context: improving decisions about point-of-use water treatment among the rural poor in Africa. Risk Anal. 32 (1), 67–80.

Avkopedia, 2010. Lifestraw family. Web, http://www.akvo.org/wiki/index.php/LifeStraw_Family(accessed October 17, 2012).

Backer, H., 2008. Water disinfection for international travelers. In: Jay, S.K., Phyllis, E.K., David, O.F., Hans, D.N., Bradley, A.C. (Eds.), Travel Medicine. second ed. Mosby, Edinburgh, pp. 47–58.

Banbury, J., 2008. A simple way to make bad water safe. In: UNICEF USA. (accessed October 28, 2009).

Bandopadhyaya, R., Sivaiah, M.V., Shankar, P.A., 2008. Silver-embedded granular activated carbon as an antibacterial medium for water purification. J. Chem. Technol. Biotechnol. 83, 1177–1180.

Bell Jr., F.A., 1991. Review of effects of silver-impregnated carbon filters on microbial water quality. J. AWWA 83 (8), 74–76.

Bell, F.A., et al., 1984. Studies on home water treatment systems. J. AWWA 76 (4), 126.

Bellamy, W.D., Silverman, G.P., Hendricks, D.W., Logsdon, G.S., 1985a. Removing Giardia cysts with SSF. J. Am. Water Works Assoc. 77 (2), 52–60.

Bellamy, W.D., Hendricks, D.W., Logsdon, G.S., 1985b. Slow sand filtration: influences of selected process variables. J. Am. Water Works Assoc. 77 (12), 62–66.

Berg, P.A., 2010. A new water treatment product for the urban poor in the developing world. In: World Environmental and Water Resources Congress 2010: Challenges of Change ASCE, pp. 2010–2025.

Bilenko, Y., Shturm, I., Bilenko, O., Shatalov, M., Gaska, R., 2010. New UV technology for point-of-use water disinfection. In: Nanotechnology 2010: Bio Sensors, Instruments,

Medical, Environment and Energy, vol. 3, pp. 601–604. Also, http://www.s-et.com/cleantech-manuscript.pdf.

Bilton, A.M., Wiesman, R., Arif, A.F.M., Zubair, S.M., Dubowsky, S., 2011. On the feasibility of community-scale photovoltaic-powered reverse osmosis desalination systems for remote locations. Renew. Energy 36, 3246–3256.

Bolton, J.R., Cotton, C.R., 2008. The Ultraviolet Disinfection Handbook. American Water Works Association, Denver, CO 168 pp.

Bowsker, C., Sain, A., Shatalov, M., Ducoste, J.J., 2011. Microbial UV fluence-response assessment using a novel UV-LED collimated beam system. Water Res. 45 (5), 2011–2019.

Brady-Estevez, A.S., Schnoor, M.H., Vecitis, C.D., Saleh, N.B., Elimelech, M., 2010. Multiwalled carbon nanotube filter: improving viral removal at low pressure. Langmuir. 26 (18), 14975–14982.

Brame, J., Li, Q., Alvarez, P.J.J., 2011. Nanotechnology-enabled water treatment and reuse: emerging opportunities and challenges for developing countries. Trends Food Sci. Tech. 22, 618–624.

Brewer, W.S., Carmichael, W.W., 1979. Microbiological characterization of granular activated carbon. J. Am. Water Works Assoc. 71, 738–740.

Burch, J.D., Thomas, K.E., 1998. An Overview of Water Disinfection in Developing Countries and the Potential for Solar Thermal Water Pasteurization. NREL/TP-550-23110. National Renewable Energy Laboratory, Golden, CO.

Buros, O.K., 2000. The ABCs of Desalting, second ed. International Desalination Association, Topsfield, MA.

Campbell, E., 2005. Study on life span of ceramic filter colloidal silver pot shaped (CFP) model. Potters for Peace.

Caslake, L.F., Connolly, D.J., Menon, V., Duncanson, C.M., Rojas, R., Tavakoli, J., 2004. Disinfection of contaminated water by using solar irradiation. Appl. Environ. Microbiol. 70 (2), 1145–1151.

Cath, T.Y., Childress, A.E., Elimelech, E., 2006. Forward osmosis: principles, applications, and recent developments. J. Membr. Sci. 281, 70–87.

Cengel, Y.A., 1998. Heat Transfer: A Practical Approach. McGraw-Hill, Manchester, UK.

Chang, J.C.H., Osoff, S.F., Lobe, D.C., Dorfman, M.H., Dumais, C.M., Qualls, R.G., Johnson, J.D., 1985. UV inactivation of pathogenic and indicator microorganisms. Appl. Environ. Microbiol. 49, 1361–1365.

Chen, Y., Worley, S.D., Kim, J., Wei, C.-I., Chen, T.-Y., Santiago, J.I., Williams, J.F., Sun, G., 2003. Biocidal poly(styrenehydantoin) beads for disinfection water. Ind. Eng. Chem. Res. 42, 280–284.

Ciochetti, D.A., Metcalf, R.H., 1984. Pasteurization of naturally contaminated water with solar energy. Appl. Environ. Microbiol. 47, 223–228.

Clasen, T., Boisson, S., 2006. Household-based ceramic water filters for the treatment of drinking water in disaster response: an assessment of a pilot programme in the Dominican Republic. Water Pract. Technol. 1 (2). http://dx.doi.org/10.2166/wpt.2006.031.

Clasen, T., Edmondson, P., 2006. Sodium dichloroisocyanurate (NaDCC) tablets as an alternative to sodium hypochlorite for the routine treatment of drinking water at the household level. Int. J. Hyg. Environ. Health 209, 173–181.

Clasen, T., Saeed, T.F., Boisson, S., Edmondson, P., Shipin, O., 2007. Household water treatment using sodium dichloroisocyanurate (NaDCC) tablets: a randomized, controlled trial to assess microbiological effectiveness in Bangladesh. Am. J. Trop. Med. Hyg. 76 (1), 187–192.

Cleary, S., 2005. Sustainable drinking water treatment for small communities using multistage slow sand filtration. M.Sc. Thesis, University of Waterloo, Waterloo, ON.

Cohen, D., 2004. Chemical Processing magazine, mixing moves osmosis technology forward. http://www.chemicalprocessing.com/articles/2004/346.html.

Colindres, R.E., Jain, S., Bowen, A., Domond, P., Mintz, E., 2007. After the flood: an evaluation of in-home drinking water treatment with combined flocculent-disinfectant following Tropical Storm Jeanne—Gonaives, Haiti, 2004. J. Water Health 5 (3), 367–374.

Crowther, K., 2010. Risk-informed assessment of regional preparedness: a case study of emergency potable water for hurricane response in Southeast Virginia. Int. J. Crit. Infrastruct. Prot. 3, 83–98.

Dalaedt, Y., Daneels, A., Declercks, P., Behets, J., Ryckeboer, J., Peters, E., Ollevier, F., 2009. E. coli and L. pneumophila. Microbiol. Res. 163, 192–199.

Dev, R., Tiwari, G.N., 2011. Solar distillation. In: Ray, C., Jain, R. (Eds.), Drinking Water Treatment: Focusing on Appropriate Technology and Sustainability. Springer, New York, 262 pp (Chapter 5).

Dies, R.W., 2003. Development of a ceramic water filter for Nepal. Master's Thesis.

Doocy, S., Burnham, G., 2006. Point-of-use water treatment and diarrhea reduction in the emergency context: an effectiveness trial in Liberia. Trop. Med. Int. Health 11 (10), 1542–1552.

Dorea, C.C., 2012. Comment on "Emergency water supply: A review of potential technologies and selection criteria" Water Res. 46 (18), 6175–6176. Available at, http://dx.doi.org/10.1016/j.watres.2012.07.062 (accessed February 19, 2013).

Driscoll, F.G., 1986. Ground Water and Wells, second ed. Johnsons Screens, St. Paul, MN, USA. 1089.

Duff, W.S., Hodgson, D., 1999. Design and evaluation of a passive flow-through solar water pasteurization system.American Solar Energy Society Conference, Portland, Maine, June.

Duff, W.S., Hodgson, D., 2005. A simple high efficiency solar water purification system. Sol. Energy 79, 25–32.

Dulbecco, R., 1949. Reactivation of ultraviolet-inactivated bacteriophage by visible light. Nature 163, 949–950.

Eckert, P., Lamberts, R., Wagner, C., 2008. The impact of climate change on drinking water supply by riverbank filtration. Water Sci. Technol. 8 (3), 319–324. http://dx.doi.org/10.2166/ws.2008.077© IWA Publishing 2008.

El-Swaify, G.A., 2013. Slow sand filtration as a low-cost water purification technology. M.S. Thesis, Department of Civil Engineering, University of Hawaii, Honolulu, HI.

EPA, 2006. Emergency disinfection of drinking water.

Feachem, R.E., Bradley, D.J., Garelick, H., Mara, D.D., 1983. Sanitation and Disease: Health Aspects Excreta and Wastewater Management. Wiley, New York.

Filtrix, 2012. Pentair. http://www.filtrix.com/EngineeredProduct_Prod_Emerge_FP-standard.aspx (accessed October 17, 2012).

Fischer, T., Day, K., Grischek, T., 2006. Sustainability of riverbank filtration in Dresden, Germany. In: UNESCO IHP-VI Series on Groundwater No. 13, Recharge Systems for Protecting and Enhancing Groundwater Resources.Proceedings of the International Symposium on Management of Artificial Recharge, June 11–16, 2005, Berlin, pp. 23–28.

Foster, S., Tuinhof, A., van Steenbergen, F., 2012. Managed groundwater development for water-supply security in Sub-Saharan Africa: investment priorities. In: International Conference on Groundwater Special Edition 2012, vol. 38, No. 3, pp. 359–366.

Galvis, G., Latorre, J., Ochoa, A.E., Visscher, J.T., 1996. Comparison of horizontal and upflow roughing filtration. In: Graham, N., Collins, R. (Eds.), Advances in Slow Sand and Alternative Biological Filtration. John Wiley & Sons Ltd., England.

Galvis, G., Latorre, J., Visscher, J.T., 1998. Multi-stage Filtration: An Innovative Water Treatment Technology. IRC International Water and Sanitation Centre. The Hague, Netherlands, TP Series, No. 34E.

Gauto, H.F., 2012. Analysis of a vertical axis wind turbine for water treatment applications. In: World Environmental and Water Resources Congress, Albuquerque, NM.

Geldreich, E.E., Taylor, R.H., Blannon, J.C., Reasoner, D.J., 1985. Bacterial colonization of point-of-use water treatment devices. J. AWWA 77 (2), 72–80.

Gelover, S., Gomez, L.A., Reyes, K., Leal, M.T., 2006. A practical demonstration of water disinfection using TiO2 films and sunlight. Water Res. 40, 3274–3280.

GE Reports, 2010. Solar-powered water purification units ship to Haiti, January 22. http://www.gereports.com/solar-powered-water-purification-units-ship-to-haiti/(accessed October 16, 2012).

Geucke, T., Deowan, S.A., Hoinkis, J., Pätzold, Ch., 2008. Performance of a small-scale RO desalinator for arsenic removal. Desalination. 239, 198–206.http://www.sciencedirect.com/science/article/pii/S0011916409000356 (accessed January 22, 2013).

Gottinger, A., McMartin, D., Price, D., Hanson, B., 2011. The effectiveness of slow sand filtration to treat Canadian rural prairie water. Can. J. Civil Eng. 38, 455–463.

Grey, D., Sadoff, C.W., 2007. Sink or swim? Water security for growth and development. Water Policy. 9, 545–571.

Grischek, T., Schoenheinz, D., Ray, C., 2002. Siting and design issues in riverbank filtration schemes. In: Ray, C., Melin, G., Linsky, R.B. (Eds.), Riverbank Filtration: Improving Source Water Quality. Kluwer Publishers, Dordrecht, The Netherlands.

Gupta, S.K., Suantio, A., Gray, A., Widyastuti, E., Jain, N., Rolos, R., Hoekstra, R.M., Quick, R., 2007. Factors associated with E. coli contamination of household drinking water among tsunami and earthquake survivors, Indonesia. Am. J. Trop. Med. Hyg. 76, 1158–1162.

Gupta, S.K., Islam, M.S., Johnston, R., Ram, P.K., Luby, S.P., 2008. The chulli water purifier: acceptability and effectiveness of an innovative strategy for household water treatment in Bangladesh. Am. J. Trop. Med. Hyg. 78 (6), 979–984.

Heijman, S.G., Rabinovitch, E., Bos, F., Olthof, N., van Dijk, J.C., 2009. Sustainable seawater desalination: stand-alone small scale windmill and reverse osmosis system. Desalination. 248, 114–117.

Heredia, M., Duffy, J., 2006. Photocatalytic destruction of water pollutants using a TiO_2 film in PET bottles. M.S. Thesis, Energy Engineering Program, University of Massachusetts Lowell, Lowell, MA.

Hill, H., Wiener, J., Warner, K., 2012. From fatalism to resilience: reducing disaster impacts through systematic investments. Disasters. 36 (2), 175–194.

Hindiyeh, M., Ali, A., 2010. Investigating the efficiency of solar energy system for drinking water disinfection. Desalination. 259, 208–215.

Hiscock, K.M., Grischek, T., 2002. Attenuation of groundwater pollution by bank filtration. J. Hydrol. 266, 141.

Horberg, L., Suter, L., Larson, T.E., 1950. Groundwater in the Peoria Region. Bulletin 39, Illinois State Water Survey, Champaign, IL.

HTI, 2010. Humanitarian water: about, HTI Water Divisions. http://www.htiwater.com/divisions/humanitarian/about.html.

Huisman, L., Wood, W., 1974. Slow sand filtration. World Health Organization, Geneva.

Idro, 2010. The portable bicycle ultra-filtration water treatment system. http://www.quantumgeos.com/archives/10_09_01/iwater_filter_system.pdf (accessed October 22, 2012).

Islam, M.F., Johnston, R.B., 2006. Household pasteurization of drinking-water: the chulli water-treatment system. J. Health Popul. Nutr. 24 (3), 356–362.

ITACA, 2005. An introduction to slow sand filtration. http://www.solutionsforwater.org/wp-content/uploads/2011/12/Slow-Sand-Filtration-Introduction-7-MB-16-Dec-2011.pdf(accessed September 2, 2013).

Jagadeesh, A., 2006. Drinking water for all. In: Tenth International Water Technology Conference.

Jagger, J., 1967. Introduction to Research in Ultraviolet Photobiology. Prentice-Hall, Inc., Englewood Cliffs, NJ.

Jain, S., Sahanoon, O.K., Blanton, E., Schmitz, A., Wannemuehler, K.A., Hoekstra, R.M., Quick, R.E., 2010. Sodium dichloroisocyanurate tablets for routine treatment of household drinking water in periurban Ghana: a randomized controlled trial. Am. J. Trop. Med. Hyg. 82, 16–22.

Jeong, J., Kim, J.Y., Cho, M., Choi, W., Yoon, J., 2007. Inactivation of Escherichia coli in the electrochemical disinfection process using a Pt anode. Chemosphere 67, 652–659.

Jeong, J., Kim, C., Yoon, J., 2009. The effect of electrode material on the generation of oxidants and microbial inactivation in the electrochemical disinfection processes. Water Res. 43, 895–901.

Jones, J., 2011. Portable nanomesh creates safer drinking water. NASA spinoff. http://spinoff.nasa.gov/Spinoff2008/er_4.html(accessed January 24, 2013).

Jorgensen, A.J., Nohr, K., Sorensen, H., Boisen, F., 1998. Decontamination of drinking water by direct heating in solar panels. J. Appl. Microbiol. 85, 441–447.

Kalisvaart, B.F., 2004. Re-use of wastewater: preventing the recovery of pathogens by using medium-pressure UV lamp technology. Water Sci. Technol. 50 (6), 337–344.

Kang, G., Roy, S., Balraj, V., 2006. Appropriate technology for rural India—solar decontamination of water for emergency settings and small communities. Trans. R. Soc. Trop. Med. Hyg. 100, 863–866.

Katadyn, 2013. Katadyn Products: desalinators. Available at, http://www.katadyn.com/usen/katadyn-products/products/katadynshopconnect/katadyn-desalinators/(accessed March 11, 2013).

Kaushal, A., Varun, 2010. Solar stills: a review. Renewable Sustainable Energy Rev. 14, 446–453.

Kee, A., 2006. The Solar Kettle-Thermos flask: a cost effective, sustainable & renewable water pasteurization system for the developing world. In: Solar Cooking and Food Processing International Conference, Granada, Spain, July 12–16, .

Kelner, A., 1949. Photoreactivation of ultraviolet-irradiated Escherichia coli, with special reference to dose-reduction principle and ultra-violet induced mutation. J. Bacteriol. 58, 511–522.

Kenissl, M., Kolbe, T., Wurtele, M., Hoa, E., 2010. Development of UV-LED disinfection, TECHNEAU. Report Within Wp2.5: Compact Units for Decentralised Water Supply, TECHNEAU, February 2010, Germany.

Kim, J., Kwon, S., Ostler, E., 2009. Antimicrobial effects of silver impregnated cellulose: potential for antimicrobial therapy. J. Biol. Eng. 3, 20–28.

Klner, A., 1949. Effect of visible light on the recovery of Streptomyces griseus candidia from ultraviolet irradiation injury. Proc. Natl. Acad. Sci. U.S.A. 35 (2), 73–79.

Kraft, A., Stadelmann, M., Blaschke, M., Kreysig, D., Sandt, B., Schoreder, F., Rennau, J., 1999a. Electrochemical water disinfection. Part I. Hypochlorite production from very dilute chloride solutions. J. Appl. Electrochem. 29, 861–868.

Kraft, A., Blaschke, M., Kreysig, D., Sandt, B., Schoreder, F., Rennau, J., 1999b. Electrochemical water disinfection. Part II: hypochlorite production from potable water, chlorine consumption and the problem of calcareous deposits. J. Appl. Electrochem. 29, 895–902.

Kraft, A., Blaschke, M., Kreysig, D., 2002. Electrochemical water disinfection Part III: hypochlorite production from potable water with ultrasound assisted cathode cleaning. J. Appl. Electrochem. 32, 597–601.

Kubare, M., Haarhoff, J., 2010. Rational design of domestic biosand filters. J. Water Supply Res. Technol. AQUA 59 (1), 1–15.

Lambert, J.H., Karvetski, C.W., Spencer, D.K., Sotirin, B.J., Liberi, D.M., Zaghloul, H.H., Koogler, J.B., Hunter, S.L., Goran, W.D., Ditmer, R.D., Linkov, I., 2012. Prioritizing infrastructure investments in Afghanistan with multiagency stakeholders and deep uncertainty of emergent conditions. J. Infrastruct. Syst. 18, 155–166.

Lantagne, D.S., Clasen, T., 2009. Point of Use Water Treatment in Emergency Response. London School of Hygiene and Tropical Medicine, London, UK.

Lantagne, D.S., Quick, R., Mintz, E.D., 2007. Household Water Treatment and Safe Storage Options in Developing Countries: A Review of Current Implementation Practices, Navigating Peace Initiative.

Lantagne, D.S., Cardinali, F., Blount, B.C., 2010. Disinfection byproduct formation and mitigation strategies in point-of-use chlorination with sodium dichloroisocyanurate in Tanzania. Am. J. Trop. Med. Hyg. 83, 135–143.

LeChevallier, M.W., Evans, T.M., Seidler, R.J., 1981. Effect of turbidity on chlorination efficiency and bacterial persistence in drinking water. Appl. Environ. Microbiol. 2, 159–167.

Lee, E.J., Schwab, K.J., 2005. Deficiencies in drinking water distribution systems in developing countries. J. Water Health 3 (2), 109–127.

Li, H., Zhu, S.X., Ni, J., 2011. Comparison of electrochemical method with ozonation, chlorination and monochloramination. Electrochim. Acta 56, 9789–9796.

Lifesaver, 2011a. Lifesaver Jerrycan Instructions Manual. Web, http://www.lifesaversystems.com/documents/jerrycan_manual.pdf(accessed October 18, 2011).

Lifesaver, 2011b. Lifesaver Jerrycan Technical Data. Web, http://www.lifesaversystems.com/documents/LIFESAVER%20Jerrycan%20Data%202011.pdf(accessed October 18, 2011).

Lifesaver, 2011c. LIFESAVER jerrycan. Web, http://www.lifesaversystems.com/lifesaver-products/lifesaver-jerrycan(accessed October 18, 2011).

Lifestraw, 2008. Verstergaard Frandsen. http://www.vestergaard-randsen.com/lifestraw/lifestraw-family and, http://www.vestergaard-frandsen.com/lifestraw/lifestraw-family/features (accessed October 17, 2012).

Linden, K.G., Shin, G.A., Sobsey, M.D., 2001. Comparative effectiveness of UV wavelengths for the inactivation of *Cryptosporidium parvum* oocysts in water. Water Sci. Technol. 43, 171–174.

Linden, K.G., Sharpless, C., Neofotistos, P., Ferran, B., Jin, S., 2004. MP lamp aging: spectral shifts and irradiance changes. In: Presented at the Annual Conference of the American Water Works Association UVDGM Stakeholder Workshop, Orlando, FL, June 2004.

Liu, C.C.K., 2009. Wind-powered reverse osmosis water desalination for Pacific Islands and remote coastal communities. Desalination and Water Purification Research and Development Program Report No. 128.

Loo, S., Fane, A.G., Krantz, W.B., Lim, T., 2012. Emergency water supply: a review of potential technologies and selection criteria. Water Res. 46, 3125–3151.

Logsdon, G.S., Kohne, R., Abel, S., LaBonde, S., 2002. Slow sand filtration for small systems. J. Environ. Eng. Sci. 1 (5), 339–348. http://dx.doi.org/10.1139/s02-025.

Lorenz, C., Windler, L., Lehmann, R.P., Schuppler, M., Von Goetz, N., Hungerbuhler, K., Heuberger, M., Nowack, B., 2012. Characterization of silver release from commercially available functional (nano)textiles. Chemosphere. 89 (7), 817–824. http://dx.doi.org/10.1016/j.chemosphere.2012.04.063.

Lougheed, T., 2006. A clear solution for dirty water. Environ. Health Perspect. 114 (7), A424–A427.

Low, C.S., 2001. Appropriate microbial indicator tests for drinking water in developing countries and assessment of ceramic water filters. Master's Thesis.

Lukhele, L.P., Krause, R.W.M., Momba, M.N.B., 2010. Synthesis of silver impregnated carbon nanotubes and cyclodextrin polyurethanes for the disinfection of water. J. Appl. Sci. 10, 65–70.

Luquer, H., 2012. Seldon Technologies, personal communication, October 23, 2012.

Makungo, R., Odiyo, J.O., Tshidzumba, N., 2011. Performance of small water treatment plants: the case study of Mutshedzi Water Treatment Plant. Phys. Chem. Earth. 36, 1151–1158.

Makwara, E.C., 2011. Water: an economic or social good? J. Soc. Dev. Afr. 26 (2), 141–163.

Malley, J.P., Ballester, N.A., Margolin, A.B., Linden, K.G., Mofidi, A., Bolton, J.R., Crozes, G., Cushing, B., Mackey, E., Laine, J.M., Janex, M.-L., 2004. Inactivation of Pathogens with Innovative UV Technologies. American Water Works Association Research Foundation, Denver, CO, USA.

Marino, M.A., Schicht, R.J., 1969. Groundwater levels and pumpage in the Peoria-Pekin area, Illinois, 1890-1966: Report of Investigation 61. Illinois State Water Survey, Champaign, IL.

Mazumdar, N., Chikindas, M.L., Uhrich, K., 2010. Slow release polymer-iodine tablets for disinfection of untreated surface water. J. Appl. Polym. Sci. 117, 329–334.

McGuigan, K.G., Conroy, R.M., Mosler, H., Preez, M., Ubomba-Jasw, E., Fernandez-Ibanez, P., 2012. Solar water disinfection (SODIS): a review from bench-top to roof-top. J. Hazard. Mater. 235–236, 29–46.

McLennan, S.D., Peterson, L.A., Rose, J.B., 2009. Comparison of point-of-use technologies for emergency disinfection of sewage-contaminated drinking water. Appl. Environ. Microbiol. 75, 7283–7286.

Medema, G.J., Juhasz-Holterman, M.H.A., Luijten, J.A., 2000. Removal of micro-organisms by bank filtration in a gravel-sand soil. In: Jülich, W., Schubew, J. (Eds.), Proceedings of the International Riverbank Filtration Conference. International Association of the Rhine Waterworks (IAWR) Amsterdam, The Netherlands, pp. 161–168.

Medentech, 2012. Aquatabs. http://www.aquatabs.com/emergency-disaster-water-purifying-tablets/making-water-safe-during-natural-disasters.html (accessed October 26, 2012).

Meierhoffer, R., Wegelin, M., 2002. Solar Water Disinfection: A Guide for the Application of SODIS. Swiss Federal Institute of Environmental Science and Technology (EAWAG), Department of Water and Sanitation in Developing Countries (SANDEC), SANDEC, Dübendorf, Switzerland.

Miller, J.E., Evans, L.R., 2006. Forward osmosis: a new approach to water purification and desalination. Sandia Report, SAND2006-4634.

Miranda, M.S., Infield, D., 2002. A wind-powered seawater reverse-osmosis system without batteries. Desalination 153, 9–16.

Mumtaz, N., Pansdey, G., Labhsetwar, P.K., Aney, S., 2012. Operating cost analysis of continuous mode electrolytic defluoridation process. Int. J. Civil Struct. Environ. Infrastruct. Eng. Res. Dev. 2 (3), 12–29.

Murphy, H.M., Sampson, M., McBean, E., Farahbakhsh, K., 2009. Influence of household practices on the performance of clay pot water filters in rural Cambodia. Desalination 248, 562–569.

NASA Spinoff, 1995. An innovation for global clean water (MCV). http://ntrs.nasa.gov/archive/nasa/casi.ntrs.nasa.gov/20020078331_2002126564.pdf, (accessed January 28, 2013), pp. 72–75.

Ngwenya, N., Ncube, E.J., Parsons, J., 2013. Recent advances in drinking water disinfection: successes and challenges. Rev. Environ. Contam. Toxicol. 222, 111–170.

Norit, 2012. Norit Waterfront (Brochure). http://www.aquasolutions.sk/data/uploads/Perfector%20en%20.pdf (accessed October 17, 2012).

Oates, P.M., Shanahan, P., Polz, M.F., 2003. Solar disinfection (SODIS): simulation of solar radiation for global assessment and application for point-of-use water treatment in Haiti. Water Res. 37, 47–54.

Oguma, K., Katayama, H., Mitani, H., Morita, S., Hirata, T., Ohgaki, S., 2001. Determination of pyrimidine dimers in *Escherichia coli* and *Cryptosporidium parvum* during UV light inactivation, photoreactivation, and dark repair. Appl. Environ. Microbiol. 67, 4630–4637.

Oguma, K., Katayama, H., Ohgaki, S., 2002. Photoreactivation of Escherichia coli after low- or medium-pressure UV disinfection determined by an endonuclease sensitive site assay. Appl. Environ. Microbiol. 68 (12), 6029–6035.

Oh, B.S., Oh, S.G., Hwang, Y.Y., Yu, H.-W., Kang, J.-W., Kim, I.S., 2010. Formation of hazardous inorganic by-products during electrolysis of seawater as a disinfection process for desalination. Sci. Total Environ. 408 (2010), 5958–5965.

Oloruntoba, R., 2010. An analysis of the Cyclone Larry emergency relief chain: some key success factors. Int. J. Prod. Econ. 126, 85–101.

Oya, A., Yoshida, S., Abe, Y., Iizuka, T., Makiyama, N., 1993. Antibacterial activated carbon fiber derived from phenolic resin containing silver nitrate. Carbon 31 (1), 71–73.

Palaez, M.A.Z., 2011. Implementing a UV disinfection system in a low-income area of Bolivia, South America. M.S. Thesis, Department of Civil & Environmental Engineering, University of Alberta, Canada.

Park, S.J., Jane, Y.S., 2003. Preparation and characterization of activated carbon fiber supported with silver metal for antibacterial behavior. J. Colloid Interface Sci. 261, 238–243.

Park, G.L., Schäfer, A.I., Richards, B.S., 2010. Renewable energy powered membrane technology: the effect of wind speed fluctuations on the performance of a wind-powered membrane system for brackish water desalination. J. Membr. Sci. 370 (1), 34–44.

Patterson, C.L., Adams, J., 2011. Emergency response planning to reduce the impact of contaminated drinking water during natural disasters. Front. Earth Sci. 5 (4), 341–349.

Pejack, E., 2011. Solar pasteurization. In: Ray, C., Jain, R. (Eds.), Drinking Water Treatment: Focusing on Appropriate Technology and Sustainability. Springer, New York, pp. 37–54.

Pejack, E., Al-Humaid, A., Al-Dossary, G., Saye, R., 1996. Simple devices for solar purification of water. In: Second World Conference on Solar Cookers, Universidad Nacional, Heridia, Costa Rica, July 12–15.

Pentair, 2012. FilterPen pocket-size membrane filtration. Retrieved from, http://www.filtrix.com/resources/images/472.pdf (accessed October 17, 2012).

Peter-varbanets, M., Zurbrügg, C., Swartz, C., Pronk, W., 2009. Decentralized systems for potable water and the potential of membrane technology. Water Res. 43, 245–265.

Poynter, S.F.B., Slade, J.S., 1977. The removal of viruses by slow sand filtration. Prog. Water Technol. 9, 75–88.

PWN Technologies, 2012. Perfector-E emergency water unit (Brochure). http://www.pwntechnologies.nl/resources/factsheets/pdf/Perfector-E%20-%20Emergency%20water%20unit.pdf(accessed October 17, 2012).

Rahaman, S., Vecitis, C.D., Elimelech, M., 2011. Electrochemical carbon-nanotube performance toward virus removal and inactivation in the presence of natural organic matter. Environ. Sci. Technol. 46, 1556–1564.

Rajagopalan, K., 2011. Membrane Desalination, Chapter 4. In: Ray, C., Jain, R. (Eds.), Drinking Water Treatment: Focusing on Appropriate Technology and Sustainability. Springer, New York.

Randtke, S.J., Horsley, M.B., 2012. Water Treatment Plant Design, fifth ed. McGraw Hill Professional, 1376 pp.

Ratelle, S., 2010. Innovative Water Technologies, Inc.'s solar-powered Sunspring™ water purification systems, Urgent Evoke. Web, http://www.urgentevoke.com/profiles/blogs/innovative-water-technologies?xg_source=activity (accessed October 16, 2012).

Rauth, A.M., 1965. The physical state of viral nucleic acid and the sensitivity of viruses to ultraviolet light. Biophys. J. 5, 257–273.

Ray, C., 2008. Worldwide potential of riverbank filtration. Clean Technol. Environ. 10 (3), 223–225.

Ray, C., Jain, R., 2011. Drinking Water Treatment: Focusing on Appropriate Technology and Sustainability. Springer, New York.

Ray, C., Borah, D.K., Soong, D., Roadcap, G.S., 1998. Agricultural chemicals: impacts on riparian municipal wells during floods. J. AWWA 90 (7), 90–100.

Ray, C., Grischek, T., Schubert, J., Wang, J.Z., Speth, T.F., 2002a. A perspective of riverbank filtration. J. AWWA 94 (4), 149–160.

Ray, C., Melin, G., Linsky, R.B., 2002b. Riverbank Filtration: Improving Source-Water Quality. Kluwer Academic Press, Dordrecht, The Netherlands.

Ray, C., Babbar, A., Yoneyama, B., Sheild, L., Respicio, B., Ishii, C., 2012. Evaluation of low cost water purification system for humanitarian assistance and disaster relief. Clean Technol. Environ. Policy 15, 345–357.

Reasoner, D.J., Blannon, J.C., Geldreich, E.E., 1987. Microbiological characteristics of third-faucet point-of-use devices. J. Am. Water Works Assoc. 79 (10), 60–66.

Regunathan, P., Beauman, W.H., 1987. Microbiological characteristics of point-of-use pre-coat carbon filters. J. AWWA 79 (10), 7–75.

Reiff, F.M., Roses, M., Venczel, L., Quick, R., Witt, V.M., 1996. Low-cost safe water for the world: a practical interim. J. Public Health Policy 17 (4), 389–408.

Rosa, G., Clasen, T., 2010. Estimating the scope of household water treatment in low- and medium-income countries. Am. J. Trop. Med. Hyg. 82 (2), 289–300.

Safapour, N., Metcalf, R., 1999. Enhancement of water pasteurization with reflectors. Appl. Environ. Microbiol. 65, 859–861.

Safe Water Systems, 2012. SunRay 1000. http://www.safewatersystems.com/sunray_1000 (accessed December 4, 2012).

Saidur, R., Elcevvadi, E.T., Mekhilef, S., Safari, A., Mohammed, H.A., 2011. An overview of different distillation methods for small scale applications. Renew. Sustain. Energy Rev. 15, 4756–4764.

Saitoh, T.S., El-Ghetany, H.H., 2002. A pilot solar water disinfecting system: performance analysis and testing. Sol. Energy 72 (3), 261–269.

Salomonsson, S., Thoresson, H., 2010. Windmill driven water pump for small-scale irrigation and domestic use: in Lake Victoria basin. Bachelor Degree Project, University of Skode.

Salsali, H., McBean, E., Brunsting, J., 2011. Virus removal of Cambodian ceramic pot water purifiers. J. Water Health 9 (2), 306–311.

Salter, R., 2006. Forward Osmosis, Water Conditioning and Purification. (white paper by R. Salter, CEO of HTI).

Sandhu, C., Grischek, T., Kumar, P., Ray, C., 2011. Potential for riverbank filtration in India. Clean Technol. Environ. Policy 13, 295–316.

Saye, R., Pejack, E., 1994. A temperature indicator for purifying water. In: Proceedings of the Second World Conference on Solar Cookers, Universidad Nacional, Heridia, Costa Rica, July 12–15.

Schicht, R.J., 1965. Ground-water development in East St. Louis Area, Illinois. Report of Investigation 51, Illinois State Water Survey, Champaign, IL.

Schijven, J.F., Hoogenboezem, W., Hassanizadeh, S.M., Peters, J.H., 1999. Modelling removal of bacteriophages MS2 and PRD1 by dune recharge at Castricum, Netherlands. Water Resour. Res. 35, 1101–1111.

Schmid, P., Kohler, M., Meierhofer, R., Luzi, S., Wegelin, M., 2008. Does the reuse of PET bottles during solar water disinfection pose a health risk due to the migration of plasticizers and other chemicals into the water? Water Res. 42, 5054–5060.

Schoenheinz, D., Grischek, T., 2010. Behaviour of dissolved organic carbon (DOC) during bank filtration under extreme climate conditions. In: Ray, C., Shamrukh, M., Ghodeif, K. (Eds.), Riverbank Filtration for Water Security in Desert Countries. Springer, Dordrecht, The Netherlands.

Schubert, J., 2002. Hydraulic aspects of riverbank filtration—field studies. J. Hydrol. 266, 152.

Schuler, P.F., Ghosh, M.M., Gopolan, P., 1991. Slow sand and diatomaceous earth filtration of cysts and other particulates. Water Res. 25 (8), 995–1005. http://dx.doi.org/10.1016/0043-1354(91)90149-K.

Seldon, 2012. Waterbox 300 MIL. http://seldontechnologies.com/products/waterbox/ (accessed October 18, 2012).

Seyedin, S.H., Jamali, H.R., 2011. Health information and communication system for emergency management in a developing country, Iran. Med. Syst. 35, 591–597.

Sharma, L., Greskowiak, J., Ray, C., Eckert, P., Prommer, H., 2012. Elucidating temperature effects on seasonal variations of biogeochemical turnover rates during riverbank filtration. J. Hydrol. 428–429, 104–115.

Sharpless, C.M., Linden, K.G., 2001. UV photolysis of nitrate: quantum yields, effects of natural organic matter and dissolved CO2, and implications for UV water disinfection. Environ. Sci. Technol. 35 (14), 2949–2955.

Skyjuice Foundation, 2010. SkyHydrant Set Up and Operating Instructions. http://www.skyjuice.com.au/documents/Instructions-SkyHydrantJun2010.pdf (accessed October 22, 2012).

Smith, E.M., Plewa, M.J., Lindell, C.L., Richardson, S.D., Mitch, W.A., 2010. Comparison of byproduct formation in waters treated with chlorine and iodine: relevance to point-of-use treatment. Envrion. Sci. Technol. 44, 8446–8452.

Solar Cookers International (SCI), 2009a. Cookit. Wikia Inc. www.solarcooking.org (accessed July 2, 2009).

Solar Cookers International (SCI), 2009b. Water Pasteurization Indicator. Wikia Inc. http://solarcooking.wikia.com/wiki/Water_Pasteurization_Indicator(accessed July 2, 2009).

Soric, A., Cesaro, R., Perez, P., Guiol, E., Moulin, P., 2012. Eausmose project desalination by reverse osmosis and batteryless solar energy: design for a 1 m3 per day delivery. Desalination. 301, 67–74.http://www.sciencedirect.com/science/article/pii/S0011916412003219 (accessed January 22, 2013).

Spectra Watermakers, 2013. Disaster relief. http://www.spectrawatermakers.com/products/all_products.html(accessed February 8, 2013).

Sphere Project, 2011. Sphere Handbook: Humanitarian Charter and Minimum Standards in Disaster Response, 2011. Available at: www.sphereproject.org/resources/download-publications/(accessed March 4, 2013).

Srinivasan, N.R., Shankar, P.A., Bandopadhyaya, R., 2013. Plasma treated activated carbon impregnated with silver nanoparticles for improved antibacterial effect in water disinfection. Carbon 57, 1–10.

Steele, A., Clarke, B., Watkins, O., 2008. Impact of jerry can disinfection in a camp environment—experiences in an IDP camp in Northern Uganda. J. Water Health 6 (4), 559–564.

Stevens, R., Johnson, R., Eckerlin, H., 1998. An investigation of a solar pasteurizer with an integral heat exchanger (SPIHX). In: American Solar Energy Society conference proceedings, 14–17 June, 1998.

Swartz, C., 2009. A planning framework to position rural water treatment in South Africa for the future. A Report to the Water Research Commission No. TT 419/09.

Taylor, R.H., Allen, M.J., Geldreich, E.E., 1979. Testing of home use carbon filters. J. AWWA 71 (10), 577–579.

Tsai, J.S., Chi, C.S.F., 2012. Cultural influence on the implementation of Incident Command System for emergency management of natural disasters. J. Homeland Secur. Emerg. Manage. 9 (1), article 29. http://dx.doi.org/10.1515/1547-7355.1970.

Tiwari, G.N., Singh, H.N., Tripathi, R., 2003. Present status of solar distillation. Sol. Energy 75, 367–373.

UNISDR, 2006. NGOs & disaster risk reduction: a preliminary review of initiatives and progress made. http://www.unisdr.org/2008/partner-netw/ngos/meeting1-october-2006/NGOs_and_DRR_Background_Paper.pdf (accessed October 31, 2012).

UNISDR, 2011. Disaster through a different lens. http://www.unisdr.org/files/20108_mediabook.pdf(accessed October 31, 2012).

UNISDR, 2012. 2011 Disasters in numbers. http://www.unisdr.org/files/24692_2011disasterstats.pdf(accessed October 31, 2012).

U.S. Agency for International Development (USAID), 1980. The USAID Desalination Manual. CH2M HILL International for the U.S. Agency for International Development, Washington, DC.

U.S. Army Public Health Command, 2010. Just the Facts. . . .Carbon Nanotubes in Drinking Water Treatment, 31-013-0410. From, http://phc.amedd.army.mil/PHC%20Resource %20Library/CarbonnanotubesApr10.pdf (accessed January 11, 2013).

USEPA, 2006. Ultraviolet disinfection guidance manual for the final long term 2 enhanced surface water treatment rule. Office of Water, EPA 815-R-R-06-007.

Valenzuela, E., Alvarez, P.E., Eapen, D., Brito, D., Camas, K., Diaz, G., Diaz, K., Espinosa, F., Vasquez, C., 2010. Water disinfection by solar radiation in a green house effect device. Int. J. Global Warming 2 (1), 48–56.

Van Halem, D., Van der Laan, H., Heijman, S.G.J., van Dijk, J.C., Amy, G.L., 2009. Assessing the sustainability of the silver-impregnated ceramic pot filter for low-cost household drinking water treatment. Phys. Chem. Earth 34, 36–42.

Vick, B.D., Neal, B.A., 2012. Analysis of off-grid hybrid wind turbine/solar PV water pumping systems. Sol. Energy 86, 1197–1207.

Visscher, J., 1990. Slow sand filtration: design, operation and maintenance. J. Am. Water Works Assoc. 82 (6), 67–71.

Vitello, M., Elmore, A.C., Crow, M., 2011. Development of a mobile water disinfection unit powered by renewable energy. J. Energy Eng. 137, 207–213.

Wang, J., 2002. Riverbank filtration case study at Louisville, Kentucky. In: Ray, C., Melin, G., Linsky, R.B. (Eds.), Riverbank Filtration: Improving Source Water Quality. Kluwer Publishers, Dordrecht, The Netherlands (Chapter 7).

Wegelin, M., 1983. Roughing filters as pre-treatment for slow sand filtration. Water Supply 1, 67–75.

Wegelin, M., Schertenleib, R., 1993. Roughing filters for water treatment. International Reference Centre for Waste Disposal, IRCWD News, No. 27, August 2003.

World Wide Water, 2013. The survival bag. http://www.worldwidewater.biz/the-survival-bag/(accessed January 28, 2013).

Wright, H., Heath, M., Schulz, C., bogan, T., Cabaj, A., Schmalwieser, A., 2009. AwwaRF UV knowledgebase documents UV disinfection in North America. In: Proceedings of the AWWA Annual Conference and Exposition, San Diego, CA, AWWA, Denver.

Yao, K.-M., Habibian, M.T., O'Melia, C.R., 1971. Water and wastewater filtration: concepts and applications. Environ. Sci. Technol. 5 (11), 1105–1112.

Yegian, D.T., Andreatta, D., 1996. Improving the performance of a solar water pasteurizer. In: Developments in Solar Cookers. In: Proceedings of the Second World Conference on Solar Cookers, Universidad Nacional, Heridia, Costa Rica, July 12–15.

Zhang, S.T., Fu, R.W., Wu, D.C., Xu, W., Ye, Q.W., Chen, Z.L., 2004. Preparation and characterization of antibacterial silver-dispersed activated carbon aerogels. Carbon. 42 (15), 3209.

Zimmer, J.L., Slawson, R.M., 2002. Potential repair of Escherichia coli DNA following exposure to UV radiation from both medium- and low-pressure UV sources used in drinking water treatment. Appl. Environ. Microbiol. 68 (7), 3293–3299.

INDEX

Note: Page numbers followed by "*f*" indicate figures and "*t*" indicate tables.

A

Arnal UF system, 124, 124*f*

B

Bank filtration (BF)
  chemical processes, 131
  description, 131, 131*f*
  long-lasting filtration technology, 130
  sand and gravel deposits, 130
Beer-Lambert law, 59
BF. *See* Bank filtration (BF)

C

Candle filter
  Clear Kisii, 103–104, 103*f*
  colloidal silver coating, 105
  cost, 106
  Hong Phuc filter, 105
  Katadyn product, 105
  Kisii system, 105
  lab tests, 104
  local construction companies, 104
  maintenance, 105
  RWD, 105
  size, 103–104, 104*f*
  water purification, 103–104
Carbon nanotube (CNT)
  iodinated resins, 172–173
  nanosorbent form, 170
  renewable energy, 170–172
  silver particles, 170
Ceramic disk filter
  colloidal silver coating, 107
  cost, 108
  description, 106, 107*f*
  local materials, 110
  maintenance, 108
  strengths and weaknesses, 110, 110*t*
  Terracotta containers, 106, 106*f*
  Thimi filter and TERAFIL, 106–107
Ceramic pot filters
  cost, 110
  description, 108, 109*f*
  local materials, 110
  maintenance, 109–110
  materials, manufacturing, removal
    efficiency, 109
  strengths and weaknesses, 110, 110*t*
  untreated *vs.* treated water product, 108,
    108*f*
Chlorinators
  Aquatab concentrations, 86, 86*t*
  chlorine tablets, 84–85
  disadvantages, 85–86
  emergency response applications, 86, 86*t*
  liquid chlorine, disinfectant, 83–84
Chlorine tablets
  Aquatab concentrations, 85, 86*t*
  diarrheal risk, 85
  diffusion-induced release, halogens, 85
  sodium dichloroisocyanurate (NaDCC),
    84–85
Chulli treatment method
  device, 115, 116*f*
  removal rate, pathogens, 115–116
  traditional clay ovens, 115
  water purifier system, 115–116
Clear Kisii filters, 103–104, 103*f*
CNT. *See* Carbon nanotube (CNT)
Conventional water treatment plants, 178
Costs
  average product cost, 15–16
  daily labor cost, 15
  description, 11
  drinking water, emergency, 15
  humanitarian relief, 11
  NGOs and programs, 16
  relative costs, 16–17
  treatment system, 11

D

Disinfection byproducts (DBPs), 158
Disinfection systems
  chlorinators, 83–86
  electrochlorination systems, 79–83

Disinfection systems *(Continued)*
  silver-impregnated activated carbon,
    74–79
  UV light systems, 55–74
Distillation technology
  brackish/sea water, 24–25
  desalinating techniques, 26
  drinking and cooking, 25
  energy-balance analysis, 26
  multistage flash plant, 25–26, 26*f*
  potable water, 25–26
  removal efficiencies, 20*t*, 24–25
  symmetric and asymmetric solar systems,
    26, 28*f*

E

Electrochlorination systems
  anodic oxidation, 82
  Cascade Design, Inc., 80, 81*f*
  cost, 82–83
  electrochemical disinfection, water,
    81–82
  household applications, 79–80
  Hydrosys AO System®, 82, 83*f*
  platinum-and iridium oxide-coated
    titanium, 82
  seawater electrolysis, bromate, 82
  small Ecodis cell, 80–81
Emergency water purification, 4

F

FilterPen
  cost, 115
  disposable device, 114–115
  small size, travelling, 114–115, 114*f*
Forward osmosis (FO)
  commercial products, 29
  costs, 29
  emergency use, 27–28
  evaluation, 29
F-specific RNA coliphages (FRNAPH),
  163, 164*f*

G

GHE. *See* Greenhouse gas effect (GHE)
Granular activated carbon (GAC) filters
  coconut GAC filters, 75
  silver on dark GAC surfaces, 78, 78*f*
  water filtration and bacterial
    endotoxins, 74
Greenhouse gas effect (GHE), 47–48

H

HA/DR. *See* Humanitarian assistance/
  disaster relief (HA/DR)
Heat exchangers
  batch-process devices, 41
  concentric tube/flat type, 43–44
  flowthrough pasteurizer, 42, 42*f*
  mass flow rate, 43
  thermostatic valves, 43
Heterotrophic plate count (HPC), 75
Humanitarian assistance/disaster relief
  (HA/DR)
  applications, 61–63
  electrochlorinator, Cascade Design, Inc.,
    80, 81*f*
  natural disasters and wars, 19–20
  RO process, 20
  sizes, UV installations, 61–63
  surface waters, 74–75
  UV disinfection, potable water, 55

I

Internally displaced person (IDP)
  camps/community, 4
  diarrhea, 1–3
  unsafe drinking water, 1–3
Iodinated resins, 172–173
iWater Cycle
  bicycle to power pumps, 128, 129*f*
  cost, 128
  typhoon victims, bicycle power to filter
    water, 128, 129*f*

L

LED. *See* Light emitting diode (LED)
Lifesaver Jerrycan
  cost, 122
  location, hand pump and ultrafiltration
    filter, 121–122, 123*f*
  testing, developing countries, 121–122,
    122*f*
  ultrafiltration (UF) membranes, 121–122

Lifestraw Family
    bacteria and virus reduction, 113
    configuration, 113–114, 113*f*
    cost, 114
    ultrafiltration-based system, 111–113,
        112*f*
Lifestraw Personal
    cleaning, 111, 112*f*
    cost, 111
    description, 111, 111*f*
    portable water treatment device, 111
Light emitting diode (LED), 69
Liquid chlorine
    advantages, 84
    degradation process, 84
    household bleach, 83–84, 84*t*
    microorganism inactivation, 83–84
    treatment device, 84
Low-pressure high-output (LPHO), 56

M
Multistage backpack filter
    cost, 118
    operating removal efficiency, 118
    system, 117, 117*f*
    and UV disinfection device, 117, 117*f*

N
Nanotechnology
    CNT, 170
    environmental effects, 169–170
    nanomaterials, 169–170
    silver particles, 170
    water treatment applications, 169–170
Natural disasters
    characteristics, 5
    climate change, 1, 2*f*
    flow rate, 5, 5*t*
    hygiene practices, 1–3
    morbidity and mortality, 1–3
    natural disasters types, 4
    point-of-use (POU) treatment, 1
    types, 4
Natural filtration
    "bank filtration" (BF) (*see* Bank
        filtration (BF))
    description, 130

design of wells (*see* Wells design)
    Flehe waterworks, vertical wells, 132,
        132*f*
    vertical/horizontal collector wells,
        131–132
Nongovernment organizations (NGOs)
    evaluation, small-scale systems, 129–130
    local/nonlocal governments, 177–178
    training/experience, 5–6
    water treatment devices, 5–6, 16

P
Packaged filtration units
    bacterial and contaminant removal, 103
    candle (*see* Candle filter)
    ceramic disk (*see* Ceramic disk filter)
    ceramic pot (*see* Ceramic pot filters)
    Chulli (Ovens) treatment (*see* Chulli
        treatment method)
    device, Lifestraw Personal (*see* Lifestraw
        Personal)
    gravity/human suction, 103
    Lifestraw Family (*see* Lifestraw Family)
    product, FilterPen, 114–115
Perfector-E water purification system
    bacteria and virus removal, 126
    cost, 127
    lifespan, UF modules, 126
    X-Flow technology, 126, 126*f*
Point-of-use (POU)
    bacteria colonization, 74
    disease and costs, 16–17
    emergency preparedness plans, 179–180
    NSF guidelines, 64
    small-scale community treatment, 1
    training, 6–7
    water treatment, 179
Pressurized filter units
    description, 116
    energy sources, 116
    Lifesaver Jerrycan, 121–122
    microfiltration (MF) and RO, 116
    multistage backpack filter (*see* Multistage
        backpack filter)
    packaged and portable RO filter
        (*see* Reverse osmosis (RO))
    WaterBox, 119–121

R

RDIC. *See* Resource Development
    International Cambodia (RDIC)
Renewable energy
    energy-recovery pumps, 171–172
    membrane desalination, 171
    power source, emergency, 170–171
    PV cells, 171
    rural irrigation, 171–172
    Solar Cube, 172
    technology devices, 171
    water purification processes, 170–171
    wind and solar energies, 172
Resource Development International
    Cambodia (RDIC), 109–110
Reverse osmosis (RO)
    cost, 119
    device costs, 21
    energy use, 20
    filters used, 118, 120*f*
    operating removal efficiency, 118–119
    packaged unit, Pelican case, 118, 119*f*
    pressure-driven membrane
        processes, 20
Rural Water Development (RWD)
    Program, 105

S

Silver-impregnated activated carbon
    antibacterial carbon fibers, 78–79
    biocide, water treatment, 74
    cartridge filtration systems, 74
    copper, antimicrobial effects, 74–75
    cost, 79
    FTIR spectra, 79
    home-use GAC filters, 74
    HPC, 75
    husk surface, filter candle, 75–78, 77*f*
    nano-silver, bacterial reduction, 75, 76*f*
    Pentek®, 75, 76*f*
    SEM image, GAC, 78, 78*f*
    Tata Swach filter and filter candle,
        75, 76*f*
    water filtration process, 75
SkyHydrant
    cost, 128
    daily cleaning, 127–128
    description, 127–128
    series, flow of treated water, 127–128,
        128*f*
    setup, 127–128, 127*f*
Slow sand filtration (SSF)
    biolayer formation, 88, 99
    coliforms and *E. coli*, 101
    community-scale/smaller systems, 91
    component, *schmutzdecke*, 92–93
    description, 88
    E. coli removal, two barrel-type, 98–99,
        98*f*
    effluent and depths, sand
        E. coli change, 101, 102*f*
        total coliform change, 101, 102*f*
        turbidity change, 101, 101*f*
    microbial community structure and
        ability, 101–103
    multistage, Pilot System 1, 93–94, 95*f*
    performance, in parallel and series, 94
    Pilot System 2, 93–94, 95*f*
    plastic tanks, community-scale, 91, 91*f*
    posttreatment devices, 101, 102*f*
    posttreatment units, 97
    raw water, 90
    removal efficiency, 89, 89*t*
    roughing filter/natural settling, 101
    set up at Fort Magsaysay, Philippines,
        92, 93*f*
    silver-impregnated activated carbon, 97
    single-barrel unit typically, household use,
        91–92, 92*f*
    stress test, two barrels, 97
    town of Falls City, Oregon, 90–91,
        90*f*
    turbidity removal
        Pilot System 1 and 2, 93–94, 96*f*
        two barrel-type, 98–99, 98*f*
    two barrels run in series
        effluent *E. coli* concentration, 100, 100*f*
        effluent turbidity, 100, 100*f*
    two filter units run parallel
        E. coli change, 99–100, 99*f*
        turbidity change, 99–100, 99*f*
    unit, water utilities, 89–90, 90*f*
    untreated fresh water, 88
    UV light, 97

Small-scale systems
  Arnal UF system, installed in
      Mozambique and Ecuador, 124,
      124*f*
  evaluation, 129–130
  iWater Cycle (*see* iWater Cycle)
  Perfector-E (*see* Perfector-E water
      purification system)
  refugee/internally displaced person (IDP)
      camps, 122–123
  SkyHydrant (*see* SkyHydrant)
  Sunspring device (*see* Sunspring device)
  treatment, 178–179
Solar cookers
  "box cooker", 33–34, 34*f*
  concentrator, 36, 36*f*
  construction and foldability, 35
  CooKit, 35–36, 35*f*
  design and operating parameters, 34
  foods, 33
  thermal insulation, 34
Solar disinfection (SODIS)
  and SOPAS (*see* Solar pasteurization
      (SOPAS))
  and titanium dioxide, 49–50
Solar pasteurization (SOPAS)
  bottle pasteurizer, 38, 39*f*
  commercial devices, 41, 41*f*
  cookers, 33–36
  devices designed, water, 36–38
  evaluation, SODIS technology, 50–51,
      52*t*
  GHE, 47–48
  indicators, water, 44–47
  1-l PET bottle, 38, 39*f*
  multi-use systems, 47
  planar reflectors, 38
  puddle pasteurizer, 38, 39*t*
  recovery heat exchange devices, 41–44
  reflectors, 40
  SODIS devices, 40
  transparent polypropylene bag, 38
  water microbiology, 31–33
SSF. *See* Slow sand filtration (SSF)
Sunspring device
  cost, 125
  description, 124

MIT kits, 125
prefilters, 125
solar panels, 124–125, 125*f*
WQA Gold Seal Program, 124–125

T
Terracotta containers, 106, 106*f*
Thermotolerant coliforms (THCOL), 163,
    164*f*
Thimi filter, 106–107
Trihalomathane (THM) precursor, 89

U
Ultraviolet (UV) light systems
  absorbance, nucleotides and nucleic acid,
      60, 61*f*
  advantages and disadvantages, 71–72
  Beer-Lambert law, 59
  bench-scale testing, 68
  biofouling prevention, 73
  characteristics, mercury vapor lamps, 56,
      57*t*
  classification, 56
  collimated beam apparatus, 65–66, 66*f*
  cost per unit water treatment, 73–74
  discharge, photons, 56
  disinfection process, 73
  DNA damage, 60
  dose and microbial inactivation, 65–66
  dose-distribution characteristics, 68–69
  dose-response curves, organisms, 68–69,
      68*f*
  electromagnetic spectrum, 56, 56*f*
  flow rates, 72
  flowthrough dynamics, 70–71
  functional system, 72–73
  HA/DR applications, 61–63
  hydraulics, reactor, 68–69
  hypothetical dose distributions, 68–69,
      68*f*
  inactivation efficiency, 60
  lamp types, 57–59, 58*f*
  LED-based lamps, 69–70
  log-inactivation values, 66–67
  maintenance, 73
  mercury vapor lamps, 56

Ultraviolet (UV) light systems *(Continued)*
  microbial action *vs.* DNA absorbance,
    60–61, 62*f*
  photoreactivation, 69
  photorepair/dark repair, 69
  power consumption, 64
  prices, 64–65
  relative response *vs.* wavelength, 60, 62*f*
  scattering, 60
  solarization, 57–59
  spectrophotometer, 59
  SteriPEN, 63–64, 63*f*
  system type, 72
  thiamine dimer formation, 60, 61*f*
  USEPA validation protocol, 66–67, 67*f*
  UVT, 59
  water purification, 69–70
Ultraviolet transmittance (UVT), 59

W
WAPI. *See* Water pasteurization indicators
  (WAPI)
WaterBox, 119–121, 121*f*
Water infrastructure development
  developed countries, 176–177
  developing countries
    emergency response phases, 177–178
    located improved water source, 177–178
    peri-urban and rural areas, 178–179
    rural areas, 179
    urban areas, 178
  government involvement
    available packaged filtration systems,
      181–182, 185*t*
    country's water resources, 181
    performance and ease of use, 182
    sustainability and social acceptability,
      182
    types, emergency filtration, 181–182,
      183*t*
    watershed management policy, 181
  investment costs, 182
  packaged filtration systems, 181–182, 185*t*
  POU and packaged technologies,
    179–180
  sanitation services and hygiene, 180–181
  structural resilience, systems, 175–176

  types, emergency filtration, 181–182,
    183*t*
  watershed management policy,
    180–181
  weather-related events, 175–176
Water pasteurization indicators (WAPI)
  boiling water, 32–33
  description, 31
  destruction, microbes, 33
  folded stainless steel, 46–47
  measurement and equipment, 31
  recording thermometer, 44
  semi-log plot, time *vs.* temperature,
    32, 32*f*
  snap disks, 45–46, 46*f*
  SODIS, 33
  SOPAS, 33, 45
  temperature, microbes, 31, 32*t*
  views, 45, 45*f*
Water quality
  parameters, disaster relief, 4, 4*t*
  quantity indicators, 4
  survival water needs, 3, 3*t*
Water treatment technologies
  acute response/long-term application, 7
  characteristics, 5
  cost, 8
  decision tree, 7, 9*f*
  disaster relief programs, 6
  emergency devices, 7
  energy source/disinfection byproducts,
    169
  flowchart, 7, 8*f*
  flow rate, 5, 5*t*
  iodinated resins, 172–173
  matrix, 7, 10*f*
  nanotechnology, 169–170
  NGOs, 5–6
  POU uptake, 6–7
  power sources, 9
  renewable energy, 170–172
  source and quality, 7
  technologies and descriptions, scores,
    9–11, 12*t*
Wells design
  atrazine concentration and collector well,
    163, 164*f*

bank filtration site, river and aquifer, 154,
    155*f*
bedrock wells, 149
changes, well radius function, 132–133,
    133*f*
characteristics, unconsolidated materials,
    147–148
climate change effects, 150–151
continuous screen slot sizes, Northern
    Gravel Company, 145, 147*t*
DBPs and TOC, 158
developing countries, 149
drawdown, 148–149
drug diclofenac and EDTA, bank
    filtration site, 163–164, 165*f*
EDTA concentration, Rhine River,
    163–164, 165*f*
floods, 151
flow paths, water pumped, 164–166, 167*f*
grain size distribution curves, Northern
    Gravel Company, 145, 146*f*
log removal, LWC collector well,
    160–161, 161*f*
log removal values, aerobic spores
    B. subtilis, 160–161, 162*f*
LWC collector well and laterals, 158, 158*f*
microorganisms/contaminants removal,
    161–162
motor mount, turbine pump, 134, 135*f*
MS-2 and PRD1 phages, Rhine River,
    163, 163*f*
natural backfill and artificial (gravel) pack,
    144–145
nonpumping water table, 144
observed and simulated temperature, 154,
    156–158, 156*f*
optimum screen entrance velocities,
    aquifer, 139, 143*t*
organic carbon, oxygen, sulfate and
    nitrate, 156–158, 157*f*
pipe outlet, submersible pump, 134, 135*f*
pipe schedules, 139, 143*t*
pretreatment, surface water, 166–167
PVC screens, screen open areas, 139, 142*t*

removal, *B. subtilis* spores, 160–161, 162*f*
riverbank filtration, 149–150
sand pumping, 147
sanitary seal, well head with submersible
    pump, 134, 135*f*
screen entrance velocity, 145–147
screen open areas, stainless steel screens,
    139, 141*t*
Sieve analysis data, heterogeneous and
    homogeneous medium, 136–139,
    137*f*, 138*f*
simulated and observed concentrations,
    oxygen, 154–158, 156*f*
Siphon tubes, pumpless wells, 149–150,
    151*f*
size distribution, materials, 136
submersible and vertical turbine pumps,
    132–134, 133*t*, 134*f*
surface waters, 150–151
temperature profile, river and collector
    well, 151–154, 153*f*
THCOL, SSRC, SOMCPH and
    FRNAPH, 163, 164*f*
THMFP, HAA6FP and TOXFP, 160,
    161*f*
TOC reduction, river water after passage,
    158–160, 159*f*, 160*f*
tubular well, sand and gravel aquifer, 144,
    145*f*
turbidity reduction through BF wells,
    151, 152*f*
two different layers, screen tailoring, 139,
    140*f*
typical gravel pack well, sand and gravel
    formations, 147–148, 148*f*
unconsolidated formations, 132–133
well development, surging, 149, 150*f*
well gravel pack material, Northern
    Gravel Company, 145, 147*t*
wells location inside bends, Elbe River,
    164–166, 166*f*

X
X-Flow technology, 126, 126*f*

Printed and bound by CPI Group (UK) Ltd, Croydon, CR0 4YY

03/10/2024

01040426-0001